OECD Development Policy Tools

Methane Abatement in Developing Countries

REGULATIONS, INCENTIVES AND FINANCE

OECD DEVELOPMENT CENTRE

Please cite this publication as:
OECD (2024), *Methane Abatement in Developing Countries: Regulations, Incentives and Finance*, OECD Development Policy Tools, OECD Publishing, Paris, https://doi.org/10.1787/f3618a78-en.

ISBN 978-92-64-56745-0 (print)
ISBN 978-92-64-62131-2 (PDF)
ISBN 978-92-64-63444-2 (HTML)
ISBN 978-92-64-82200-9 (epub)

OECD Development Policy Tools
ISSN 2518-6248 (print)
ISSN 2518-3702 (online)

Photo credits: Cover design by Aida Buendía using images from © Yusnizam Yusof/Shutterstock.com; © noomcpkProvider/Shutterstock.com.

Corrigenda to OECD publications may be found at: https://www.oecd.org/en/publications/support/corrigenda.html.

Foreword

At its Fifth High-Level Meeting (HLM) held on 21 May 2019, the Members of the Governing Board invited the OECD Development Centre to "help design transformational development strategies aligned with the 2030 Agenda focusing on the sustainable transition of natural resource-rich developing countries towards a low-carbon economy and better integration into global value chains". In response to this request the Centre published in 2022 the *Equitable Framework and Finance for Extractive-based Countries in Transition (EFFECT)* and launched it at COP27. EFFECT provides a toolbox with policy options for resource-rich countries to manage the "transition away from fossil fuels in energy systems, in a just, orderly and equitable manner, accelerating action in this critical decade, so as to achieve net zero by 2050" as called for in the COP28 Global Stocktake Decision.

This report is an output of the implementation phase of EFFECT, carried out in close collaboration with the African Development Bank (AfDB). An open, intense, and enriching multi-stakeholder consultation process with natural gas producing and consuming economies, industry, international organisations, development finance institutions and non-governmental organisations participating in the EFFECT Community of Practice on Just Transition Pathways for Oil & Gas Producing Developing Countries informed the development of the report and its recommendations. An initial draft was prepared by the Development Centre and discussed during the third call of the Community of Practice on 24 October 2023. Revised drafts were presented at the second call of the Advisory Group on EFFECT implementation on 29 November 2023 and the fourth call of the Community of Practice on 19 March 2024. An advanced draft was presented at the combined teleconference of the Community of Practice and the Advisory Group on 25 June 2024. A final draft was discussed during the sixth call of the Community of Practice on 2 October 2024.

This report provides recommendations on the design of robust regulatory frameworks on methane abatement in the upstream oil and gas sector, as a crucial component of broader efforts to achieve accelerated methane emissions reduction. It sets out the enabling conditions as well as the incentives and financing mechanisms for the deployment of cost-effective methane abatement solutions in developing countries producing oil and gas. As a result, this report supports developing countries to "accelerate and substantially reduce non-carbon-dioxide emissions globally, including in particular methane emissions by 2030", as agreed at COP28, by fostering alignment between their development, climate, and energy agendas. This report is also intended to support engagement with developing countries producing oil and gas and upstream actors in the natural gas value chain for the operationalisation of emerging international initiatives – such as the European Union's Methane Abatement Partnership Roadmap and the Coalition for LNG Emission Abatement toward Net-zero ("CLEAN") initiative, spearheaded by Japan and Korea – to clean up the value chain of internationally traded fossil fuels and move from voluntary commitments to action to drive down methane emissions. Lastly, this report supports G7 efforts "to work with non G7 producing countries to reduce the methane emission intensity of imported fuels", by offering options "for regulatory approaches and market-based instruments to support methane emission reduction actions" as called for in the 2024 G7 Climate, Energy and Environment Ministers' Meeting Communiqué.

Acknowledgements

This report was prepared by Elliot Smith, Legal Analyst, Natural Resources for Development Unit, OECD Development Centre, and Lahra Liberti, Head of Unit. Lahra Liberti provided overall direction, guidance, and careful review during the drafting process. This report was developed within the framework of the Centre's Policy Dialogue on Natural Resource-based Development, as an implementing tool of the OECD *Equitable Framework and Finance for Extractive-based Countries in Transition (EFFECT)*.

The authors wish to express their deep gratitude to Brendan Devlin, former Strategy and Foresight Counsellor, Directorate General for Energy, European Commission and Dr Kelechi O. Ofoegbu, Senior Technical Adviser to the Honourable Minister, Ministry of Petroleum Resources, Federal Republic of Nigeria, for their leadership and guidance as co-chairs of the EFFECT Community of Practice on Just Transition Pathways for Oil & Gas Producing Developing Countries and for their continuous support over two years. The authors are also grateful to Ambassador Tibor Stelbaczky, Principal Adviser on Energy Diplomacy, European External Action Service (EEAS) for his leadership and overall strategic guidance as the chair of the Advisory Group on EFFECT implementation.

The authors wish to acknowledge the valuable inputs received during targeted consultation periods in November 2023, and July and October 2024, including representatives from the African Development Bank (AfDB); Asian Development Bank (ADB); Clean Air Task Force (CATF); Climate and Clean Air Coalition (CCAC); Directorate General for Energy, European Commission; Ministry of Petroleum and Mineral Resources, Egypt; Global Flaring and Methane Reduction Partnership (GFMR); Matesco & Partners; Oil and Gas Climate Initiative (OGCI); and Shell plc. In addition, the authors are grateful for the active engagement and insights from Cheikh Brahim Haiballa, *Conseiller à la Direction générale, Société Mauritanienne des Hydrocarbures* (SMH); Manfredi Caltagirone, Head, International Methane Emissions Observatory (IMEO), United Nations Environment Programme (UNEP); Meghan Demeter, Engagement & Communications Lead and Government Engagement Manager, IMEO, UNEP; Antonio Hill, Advisor, Natural Resources Governance Institute (NRGI); Dr Valerie Marcel, Executive Director, New Producers for Sustainable Energy; Christian Schwarck, Senior Policy & Advocacy Advisor Upstream, Shell; and for the expert contributions from Gergely Molnar, Energy Analyst, Natural Gas, Directorate of Energy Markets and Security, of the International Energy Agency (IEA).

The authors are also grateful for the insightful comments and valuable input received from colleagues in the OECD's Environment Directorate, in particular Joseph Cordonnier, Deger Saygin and Cécile Seguineaud.

The authors also wish to express their appreciation and gratitude to all the members of the EFFECT Community of Practice on Just Transition Pathways for Oil & Gas Producing Developing Countries, in particular for the contributions received during the Third call (24 October 2023), Fourth call (19 March 2024), as well as input received during the Second call of the Advisory Group on EFFECT implementation (29 November 2023), the combined teleconference of the Community of Practice and the Advisory Group (25 June 2024), and the sixth call of the Community of Practice on 2 October 2024.

The authors wish to thank the high-level panellists who participated in the AfDB/OECD COP28 Virtual Panel *Renewables and Natural Gas: Friends or Foes of a Just Transition in Developing Countries?*, and whose contributions informed the development of the report: Dr Rashid Ali Abdallah, Director, African Union's Africa Energy Commission (AFREC); H.E. Cristina Duarte, United Nations Under-Secretary-General and Special Adviser on Africa; Christophe McGlade, Head of the Energy Supply Unit, International Energy Agency (IEA); Jarredine Morris, Co-Head at Carbon Trust Africa; and Charles Nyirahuku, Chief Gas Regulatory Officer, African Natural Resources Management and Investment Centre (ANRC), African Development Bank (AfDB).

Lastly, the authors are grateful to the Development Centre's Communications team, Aida Buendía, Delphine Grandrieux, Elizabeth Nash, and Henri-Bernard Solignac-Lecomte for their support in the preparation of the report.

Table of contents

Tables

Figures

Boxes

Executive summary

The global oil and gas sector is the second highest source of anthropogenic methane emissions after agriculture, but it also has the largest abatement potential. Yet, global methane emissions are not reducing at the scale and pace needed to limit warming to a level consistent with Paris-aligned 1.5°C pathways, as global demand for natural gas is growing and expected to reach new all-time highs in 2024 and 2025 according to IEA projections. In the absence of concerted, Paris-aligned action on climate, current government plans and projections would lead to an increase in global oil and gas production until at least 2050 according to the 2023 Production Gap Report. At a time of global energy and climate crises, methane abatement has become an imperative for climate mitigation efforts consistent with the objective of achieving global net-zero energy systems by mid-century but also for meeting pressing energy security needs. Developing countries producing oil and gas stand to benefit in several respects from the adoption of upstream oil and gas methane abatement regulations:

- There is growing evidence that the introduction of methane regulations can lead to significant emissions reduction in the upstream segment of the value chain.
- Methane emissions regulations can help developing countries producing oil and gas improve public health and air quality, and retain the competitiveness of their exports as natural gas importing requirements tighten. They are also a key enabler for the effective implementation of emerging cooperative frameworks for cleaning up internationally traded fossil fuels.
- Regulatory requirements consistent with international standards on measurement, monitoring, reporting and verification can create a level playing field and a conducive enabling environment across public and private actors in the upstream value chain to avoid shifting emissions to developing countries and the proliferation of voluntary standards. International harmonisation supports enhanced international coherence in climate mitigation, enables cost-effective corporate compliance across jurisdictions, and reduces green washing concerns.

To that end, developing countries producing oil and gas are advised to develop robust methane abatement frameworks with the following elements:

- *Set specific, and ideally incremental, sector-specific methane emissions reduction targets* in Nationally Determined Contributions (NDCs), and embed these within Long-term Low Greenhouse Gas Emission Development Strategies (LT-LEDS), by clearly articulating how natural gas production and use contributes to the achievement of climate goals, sustainable development priorities and energy security needs. This will send clear signals to investors, help mobilise financial support, and contribute to broader systemic transformation toward net-zero energy systems.
- *Build national inventories and baselines and set methane measurement, monitoring, reporting, and verification* (MMRV) *requirements* consistent with international reporting standards (such as the UNEP's Oil & Gas Methane Partnership 2.0).
- *Set out prescriptive or performance-based equipment and technology standards or a combination of both, and design robust leak detection and repair programmes* (LDAR). Regulatory frameworks

should require operators to undertake leak detection campaigns with specified frequency, detection thresholds and time limits for repairs.

- *Put in place policies and regulations to eliminate routine flaring* (i.e. incomplete combustion of natural gas also releases methane) *and venting of natural gas* (i.e. the intentional release of gas into the atmosphere) and set out exceptional circumstances for non-routine flaring with sufficiently high flare efficiency ratios. Companies should satisfy the regulatory authority that they have investigated all reasonable alternatives to non-routine flaring and venting, before being granted permission to flare or vent associated gas for economic reasons.
- *Establish a carefully balanced reward-penalty system* to make it cost-effective for public and private companies to reduce upstream methane emissions, and clearly allocate costs associated with methane abatement to encourage compliance. The reward component can take different forms, including direct public financial support for abatement, cost-recovery provisions under production sharing agreements (PSAs), or the generation of credits to be sold on the open market. These should be combined with financial charges, such as fees, taxes or penalties to penalise companies that fail to comply with regulations. These may be in the form of royalty obligations for exceeding allowable flaring volumes, or denial of drilling permits if the capture targets of associated gas – an often unwanted product of gas production – are not achieved. To avoid wasteful gas flaring, enabling factors include the disclosure of venting and flaring profiles, granting third-party access to existing gas networks as well as preferential market access to the gas network and wholesale market. However, associated gas monetisation may require the development of transport (i.e. pipelines and distribution networks) and energy infrastructure (i.e. processing facilities and gas-fired power plants), carrying not only significant CAPEX, but also transition risks, in terms of potential gas lock-in and stranded assets, which may deter financing. A few multilateral development banks provide financing for midstream and downstream gas infrastructure projects provided investments are aligned with a country's climate policy; do not create a risk for carbon lock-in or stranded assets; reduce the energy sector's carbon intensity; and utilise best available technologies and sector best practices in limiting methane emissions.
- *Introduce methane reduction requirements at the licensing and planning stage* of any new abated oil and gas projects where still necessary in the energy transition to ensure their alignment with climate objectives.

However, upstream methane abatement finance remains well below needs. Industry finance, through compliance funds or special purpose vehicles; emissions trading schemes and sustainability-linked financing (i.e. transition and sustainability-linked bonds); or debt-for-climate swaps can help often largely indebted or fiscally constrained developing countries accelerate methane emission reductions. Sovereign and strategic investment funds represent potential additional sources of financing with no or fewer restrictions with regard to investing in the fossil fuel sector.

1 Introduction

Since the beginning of the industrial revolution, methane is estimated to be responsible for at least 25% of the rise in global temperatures. Despite representing a smaller contribution than carbon dioxide thus far, the warming potential of methane in its first 20 years in the atmosphere is more than 80 times greater than that of carbon dioxide (Cahill and Dawes, 2023[1]).

In 2023, the global oil and gas sector was responsible for over 78 million tonnes of methane emissions, representing the second highest source of anthropogenic emissions after agriculture, and ahead of waste, coal, bioenergy and biomass burning – see Figure 1.1 (IEA, 2024[2]). Significant reductions are needed across all types of greenhouse gas (GHG) emissions to achieve the Paris Agreement's temperature goal (Jeudy-Hugo, 2024[3]). While there is an urgent need to tackle all forms of anthropogenic emissions, there is an immediate opportunity to address methane emissions which need to be cut by 33% globally by 2030. In this regard, the energy sector has the largest abatement potential, given the possibility to achieve quick wins with available cost-effective abatement solutions. Indeed, the International Energy Agency (IEA) has estimated that around 70% of methane emissions from fossil fuel operations could be reduced with existing technology, and about 40% of those emissions at no net cost because the outlays for the abatement measures are less than the market value of the additional gas that is captured (IEA, 2023[4]).

Figure 1.1. Global anthropogenic methane emissions by sector in 2022

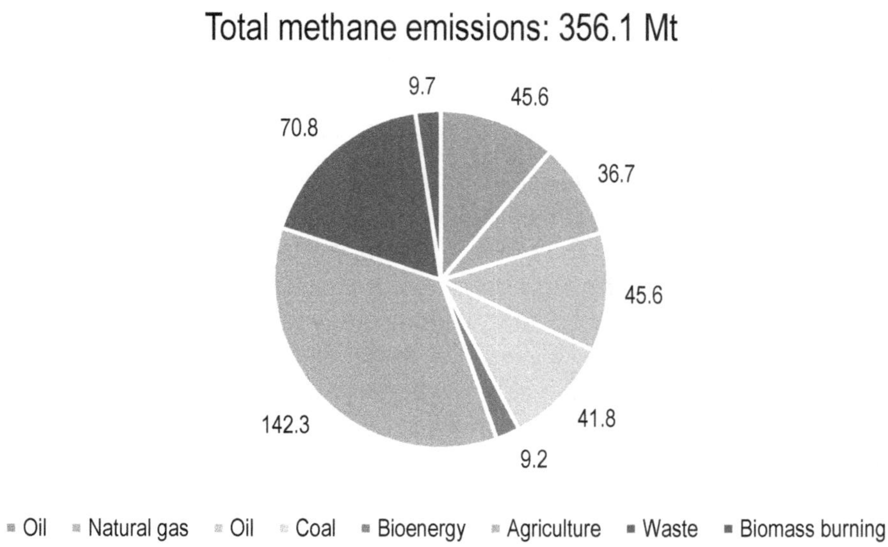

Source: (IEA, 2023[4]).

The urgency to tackle methane emissions from the oil and gas sector is also underpinned by a recognition that oil and gas production will not cease in the short-medium term. While the managed decline of oil and gas within the global energy mix is necessary to meet the goals of the Paris Agreement and to prevent irreversible damage to the environment and ecosystems, data from the 2023 Production Gap Report notes that current government plans and projections would lead to an increase in global oil and gas production until at least 2050, resulting in more than double the amount of fossil fuels in 2030 that would be consistent with limiting global warming to 1.5°C (SEI, Climate Analytics, E3G, IISD, UNEP, 2023[5]). Lastly, even under the IEA's *Net Zero Roadmap* fossil fuels would still represent a 20% share of the global energy supply in 2050 (IEA, 2021[6]). Many developing countries have also outlined plans to utilise their reserves of natural gas as a transition fuel. Consequently, the rapid reduction of methane emissions in the oil and gas sector is a crucial component of global decarbonisation efforts. Furthermore, in addition to meeting climate objectives, the reduction of methane emissions can also contribute to improved health and environmental conditions, by improving air quality especially for communities and workers around oil and gas infrastructure, reducing the rates of asthma and lessening the severity of extreme weather events. According to a recent study, targeted methane abatement in the fossil fuel sector can prevent nearly 1 million premature deaths due to ozone exposure, 90 million tonnes of crop losses due to ozone and climate changes, and about 85 billion hours of lost labour due to heat exposure by 2050 (IEA, 2023[7]).

Increasing global recognition of the urgency of tackling fossil methane emissions

At a time of global energy and climate crises, the role of natural gas in the low-carbon transition and the importance of methane abatement for enhanced energy security and climate mitigation efforts have become all the more critical for the world's top natural gas importing countries following Russia's invasion of Ukraine in February 2022.

To balance energy security concerns and climate objectives, a number of voluntary international initiatives were launched to drive collective global action to tackle oil and gas sectors methane emissions. Back in 2016, the United States, Canada, and Mexico created the North American Climate, Clean Energy, and Environment Partnership to reduce methane emissions from the oil and gas sector by 40-45% by 2025 (compared with 2012 levels) by adopting methane regulations, and by urging other G20 members to make similar commitments (The White House, 2016[8]). At COP26, the Global Methane Pledge (GMP) was launched with the ambition to reduce global methane emissions by 30% from 2020 levels by 2030. Since then, 155 countries have committed to develop national methane action plans to drive methane reductions in key methane-emitting sectors – particularly in oil and gas (Global Methane Pledge, 2021[9]).

Recognising the urgent need to reduce emissions from fossil energy value chains, Canada, the European Union, Japan, Norway, Singapore, the United Kingdom, and the United States signed up to the *Joint Declaration from Energy Importers and Exporters on Reducing Greenhouse Gas Emissions from Fossil Fuels* at COP27. The signatories to the Declaration are committed to working towards an international market for fossil energy that reduces combustion, methane and carbon dioxide emissions across the value chain to the extent feasible, and also works to reduce consumption of fossil fuels. This includes domestic an international action to: 1) adopt policies and measures to eliminate routine venting and flaring and conduct regular leak detection and repair campaigns in oil and gas exploration, production and downstream operations; 2) adopt policies and measures to support robust measurement, monitoring, reporting and verification of methane emissions in the fossil energy sector and to ensure transparency of related data; 3) mobilise technical assistance and financing to mitigate methane and carbon dioxide emissions in the fossil energy sector. In this regard in the lead up to COP28, the United States launched the Methane Finance Sprint to mobilise private, public and philanthropic finance to support developing countries to cut their methane emissions in line with the GMP.

Building on the Declaration, Japan and Korea spearheaded the Coalition for LNG Emission Abatement toward Net-zero (CLEAN) initiative, whereas the European Union announced the Methane Abatement Partnerships (MAP) Roadmap.

The CLEAN initiative was launched in 2023 by JERA Co., Inc. (JERA) and Korea Gas Corporation (KOGAS), two leading LNG importers from Japan and Korea, and the Japan Organization for Metals and Energy Security (JOGMEC), a Japanese governmental agency that seeks to ensure a stable energy supply and accelerate progress on methane emissions mitigation. CLEAN aims to broaden engagement with other LNG buyers and producers, foster collaboration between LNG producers and consumers on methane abatement solutions, increase transparency on methane emissions across the LNG value chain, and disseminate best practices on mitigation efforts along the LNG value chain. Developing LNG producers could benefit from JOGMEC's support for accelerated methane measurement and mitigation, leveraging ongoing technical collaboration with Petronas and Pertamina on emission assessment and reduction in Australia, Indonesia and Malaysia.

As announced by President von der Leyen at COP28, the European Commission will launch a Methane Abatement Partnership (MAP) Roadmap (formerly "You Collect We Buy") at COP29 to accelerate the reduction of methane emissions associated with fossil energy production and consumption. The Roadmap provides a globally adaptable, step-by step blueprint for the implementation of importer-exporter partnerships, and sets out several pillars for a co-operation framework, including a robust monitoring, reporting and verification (MRV) system built on the Oil and Gas Methane Partnership 2.0 (OGMP) framework. This is complemented by other relevant measures and policies, as well a project plan on the timeline, abatement targets, expenditure, investments and available tools in co-operation with organisations, such as the IEA, United Nations Environment Programme (UNEP's) International Methane Emission Observatory (IMEO), OECD, and Climate and Clean Air Coalition (CCAC). The Roadmap aims to mobilise efforts under the GMP, incentivise importer-exporter co-operation in support of companies improving their MRV abilities to mitigate methane emissions, and attract private investments, while contributing to both decarbonising energy systems but also ensuring security of supply. First implementation examples could be showcased at COP30.

Finally, a number of voluntary industry-led initiatives have also emerged in recent years to accelerate climate action within the oil and gas industry, including the Methane Guiding Principles, the Oil and Gas Climate Initiative (OGCI), the Oil and Gas Methane Partnership 2.0 (OGMP), and the Oil and Gas Decarbonization Charter where participating companies, including international oil companies (IOCs) and national oil companies (NOCs) committed to near-zero upstream methane emissions, ending routine flaring by 2030, and achieving net-zero operations by 2050 (UAE COP Presidency, 2023[10]), lowering methane emissions intensity and increased transparency around their emissions. Industry-led initiatives draw on the relative strengths of companies (as opposed to governments) to tackle emissions, as they have the technical capabilities to manage methane emissions, are closer to the problem at hand, and can often respond more quickly than governments when required (IEA, 2023[4]).

Why oil and gas producing countries need to take urgent action to regulate methane emissions

Voluntary approaches are not delivering sufficient methane emissions cuts

Despite the proliferation of different voluntary initiatives at all levels, global methane emissions are not reducing at the scale and pace needed to limit warming to a level consistent with Paris-aligned 1.5°C pathways. Large methane emissions events detected by satellites rose by more than 50% in 2023 compared with 2022, and recent research from the data analytics company Kayrros shows that signatories to the GMP are not on track to cut emissions by 30% by 2030 as there has been no overall reduction in the methane emissions by the majority of signatories (IEA, 2024[2]; Kayrros, 2023[11]). In addition, while the objectives of the GMP and other voluntary initiatives are global in nature, the existence of domestic policies and measures, including enabling regulations to reduce methane emissions, remains limited. A 2023 study of 281 global methane mitigation policies found that 90% of these policies have been enacted in only three regions: Asia-Pacific, Europe and North America – representing only 17% of global methane emissions (Olczak, Piebalgs and Balcombe, 2023[12]). In its 2024 Methane Tracker, the IEA estimates that if all methane policies and pledges made by countries and companies to date are implemented and achieved in full and on time, methane emissions from fossil fuels would decline by around 50% by 2030. However, in most cases, these pledges are not yet backed up by detailed plans, policies and regulations. The detailed methane policies and regulations that currently exist would cut emissions from fossil fuel operations by around 20% from 2023 levels by 2030. This falls significantly short of putting the global energy sector on track to achieve net zero emissions by mid-century, which would require reducing methane emissions from fossil fuel operations by around 75% by 2030 (IEA, 2024[2]).

There is growing international evidence that methane emissions are significantly underreported. Government inventories of methane emissions, including those submitted under the United Nations Framework Convention on Climate Change (UNFCCC) national reporting process, rarely make use of direct measurements. Instead, they are ordinarily estimated via bottom-up approaches where a typical emissions factor is applied to the total number of oil and components (e.g. wellheads, pneumatic devices, storage tanks) to estimate total emissions. However, there is a significant discrepancy between bottom-up inventory estimates and top-down methods which measure atmospheric methane concentrations using remote sensing (satellites and airborne instruments), with data from the IEA suggesting that global methane emissions from the energy sector are around 70% greater than the amount national governments have officially reported (IEA, 2023[4]). There are several explanations for this discrepancy. Governments do not submit GHG data to the UNFCCC regularly and many submissions are based on out-of-date or and incomplete picture of emissions. In addition, super-emitting events that produce a disproportionate share of total methane emissions are often not included in bottom-up inventories (Cahill and Dawes, 2023[1]).

Underreporting of methane emissions is not limited to governments. Recent UNEP data notes a significant difference between industry reported emissions and global measurement-based assessments. An analysis of company-reported emissions of the 92 member companies of the OGMP 2.0, collectively representing 34% of global oil and gas production, identified a significant disparity between their company-reported emissions and what would be expected based on atmospheric measurements (UNEP, 2023[13]). Furthermore, if companies that report emissions to the OGMP 2.0 were representative of the industry globally, this would imply that global oil and gas methane emissions in 2023 were around 95% lower than estimated by the IEA (IEA, 2024[2]).

Lastly, global energy demand is growing, and this could significantly increase methane emissions if part of the demand is met by increased production of oil and gas. While global gas demand rose just 0.5% in 2023, projections for 2024 forecast 2.5% growth primarily driven by the industrial and power sectors in fast-growing economies in Asia and gas-rich countries in Africa and the Middle East (IEA, 2024[14]). Rapid industrialisation coupled with demographic pressure and broader development objectives will result in an increase of methane emissions across oil and gas developing countries if left unchecked.

Maintaining competitiveness and preserving market access

Oil and gas developing producing countries are directly exposed to emerging international initiatives to clean up the natural gas value chain, including through new regulatory requirements imposed on their exports of oil and gas by importing jurisdictions. For example, on 10 April 2024, the European Union, the world's largest importer of fossil energy, adopted a regulation which will require that from 2027 imported oil and gas are subject to MRV requirements at the level of the producer that are equivalent to those set out in the regulation. From 2030 a methane intensity standard will also apply on imported oil and gas. Developing producer countries who are unable to meet these new MRV and methane intensity requirements will face challenges with maintaining market access or may be forced to sell oil and gas at a significant discount.

Box 1.1. EU methane regulations – provisions on imported oil and gas

In May 2024, the European Union adopted a new Regulation on methane emissions reduction in the energy sector . The regulations impose new requirements on imports of oil and gas from third countries. Specifically:

- **Monitoring, reporting and verification measures** – from 1 January 2027, importers shall demonstrate that the contracts concluded or renewed after the entry into force of the Regulation cover solely crude oil or natural gas subject to monitoring, reporting and verification measures, at the level of the producer, equivalent to those set out in the EU regulation.
- **Methane intensity** – from 2027, the EU will adopt a delegated act to determine a methodology for maximum emissions intensities for imports. From 2030, importers will have to demonstrate to EU regulators that the methane intensity associated to the crude oil or natural gas that they import is below the maximum methane intensity threshold.
- **Methane transparency database and methane performance profiles** – from February 2026, the Commission shall establish a methane transparency database, including at least data on third countries, undertakings, and importers.
- **Methane emitters global monitoring tool and rapid reaction mechanism** – By 2025, the EU shall establish a global methane monitoring tool based on satellite data and input from several certified data providers and services, including the Copernicus component of the EU Space Programme.

Source: (European Commission, 2023[15]).

Market access challenges may be compounded should other major importing jurisdictions follow suit and impose similar regulatory requirements. This is a plausible development considering that signatories to the *Joint Declaration from Energy Importers and Exporters on Reducing Greenhouse Gas Emissions from Fossil Fuels*, have already agreed to support domestic and international action to achieve emissions reductions across the fossil fuel value chain, including by "[p]utting in place measures to require or strongly incentivise reductions in greenhouse gas emissions associated with fossil energy imports". This shows how the environmental credentials of natural gas, including liquified natural gas (LNG), are coming under increasing scrutiny. In fact, despite emitting about half the CO_2 of coal when combusted, the LNG value chain remains carbon intensive and is subject to significant methane leakage (Di Odoardo et al., 2024[16]). Producer governments should also anticipate that several of the largest LNG importers (e.g. Japan, Korea, and the United Kingdom) have pledged to become net-zero by 2050, and by 2060 in China's case (OECD, 2022[17]).

Ensuring a level playing field across public and private actors

Governments should seek to create a level-playing field for methane abatement and reporting requirements across importing and exporting jurisdictions. If carefully designed, the adoption of government policies and regulations on methane emissions reduction can create an enabling conducive environment, by clarifying business expectations and providing the right incentives for IOCs and NOCs to work together to mitigate methane emissions across their operations. Governments can draw on international best practice reviewed in this report to inform the design of new methane regulations, while accounting for specific country contexts and capabilities. This will support international coherence and co-ordination for effective methane abatement and can also help align investment and financial flows.

Methane emissions requirements consistent with international standards can help achieve coherence both domestically across oil and gas operations and internationally, thus reducing compliance costs for businesses, and countering risks of underreporting. Ensuring a level-playing field for methane abatement and reporting requirements across importing and exporting jurisdictions will also address the risk of unfair climate competition (where a company in one jurisdiction is subject to additional compliance costs, whereas in other jurisdictions it is not) and can mitigate the risk of shifting emissions to developing countries.

Methane regulations have led to emissions reductions

There is growing evidence that the introduction of methane regulations can lead to significant methane emissions reduction. For example, in 2014, the province of Alberta (Canada) implemented regulations to limit methane emissions from heavy oil production facilities in the Peace River region, including requirements to eliminate routine venting, limit non-routine flaring, and conduct monthly leak detection surveys at high-risk sources. A 2022 study showed that these regulatory measures had reduced methane emissions from approximately three million cubic metres to a near-zero level over a five-year period. Furthermore, these reductions were achieved without significantly impacting production as oil production from heavy oil production facilities declined by a similar amount in both the province (41%) and Peace River region (45%), while venting in Peace River decreased by 100% (Connoy, McKenzie and Gorski, 2022[18]).

A 2023 study of methane emissions in the United States' Permian Basin showed that Texas saw twice as many methane leaks as New Mexico between 2019 and October 2023 despite sharing the same geology and many of the same operators. This difference has been attributed to the introduction of methane emissions regulations in New Mexico in 2021 to reduce routine methane venting and flaring, alongside increased compliance activities, particularly in respect of smaller facilities (Kayrros, 2023[11]); (Reuters, 2023[19]). Further evidence of the effectiveness of methane regulations can be seen in the implementation of LDAR measures. A study of thirty-six facilities in the United States and Canada found that methane emissions decreased by 44% following the successful application of a LDAR survey (Ravikumar, 2020[20]), and in California, LDAR regulations reduced fugitive emissions by up to 9% (Tran et al., 2022[21]).

Why focus on methane emissions from the upstream oil and gas sector?

Methane emissions arise from processes and equipment across the different segments of the oil and gas value chain – from exploration drilling through to delivery to the consumer. They are either the result of fugitive emissions (leaks and other irregular releases from equipment), or the intentional flaring and venting of methane, usually as a by-product of oil production (IGSD, 2023[22]). The oil and gas value chain can be categorised as follows:

- **Upstream** – oil and gas wells (both onshore and offshore), oil separation facilities, gas processing facilities, and short-distance pipelines connecting these facilities
- **Midstream** – transportation infrastructure, including pipelines and associated compressor stations, liquefied natural gas facilities and tankers, and storage facilities
- **Downstream** – distribution networks designed to reach end-consumers, including residential, commercial, and industrial (CLDP, 2023[23]).

See Table 1.1 for an overview of where sources of methane emissions arise.

Table 1.1. Sources of emissions in the oil and gas value chain

	Upstream			Midstream		Downstream
Phase	*Exploration and pre-production*	*Production*	*Gathering and processing*	*Transmission*	*Storage*	*Distribution*
Source of emissions	Site preparation	Pneumatic controllers	Compressor venting	Compressor venting and leaks	Compressors	Pipe leakages
	Drilling	Liquids unloading	Pneumatics venting leaks	Pneumatic controllers	Pneumatic controllers	Metering and regulating station leaks and vents
	Well completion	Workovers	Liquid storage tanks	Pipe leakages	Fugitives	
	Fracking (unconventional)	Flared gas	Flared gas	Storage site venting and leaks		
		Fugitives	Vented gas			

Note: Table excludes specific emissions from LNG.
Source: Adapted from (Balcombe, Brandon and Hawkes, 2018[25]); (Balcombe et al., 2015[26]).

The amount of methane emitted from operations across the oil and gas value chain is extensive. According to the IEA, the total methane lost in 2023 from global fossil fuel operations was 170 billion cubic metres, more than the natural gas production of Qatar (IEA, 2024[2]). While methane emissions arise across the different segments of the oil and gas value chain, it is the upstream segment of the value chain where emissions are most prominent and where oil and gas producing countries can have the greatest impact in reducing emissions – see Figure 1.2 (IEA, 2023[4]). For this reason, this report focuses on methane emissions from upstream oil and gas production as they can achieve deep emissions cuts.

Figure 1.2. Estimated global methane emissions by value chain segment

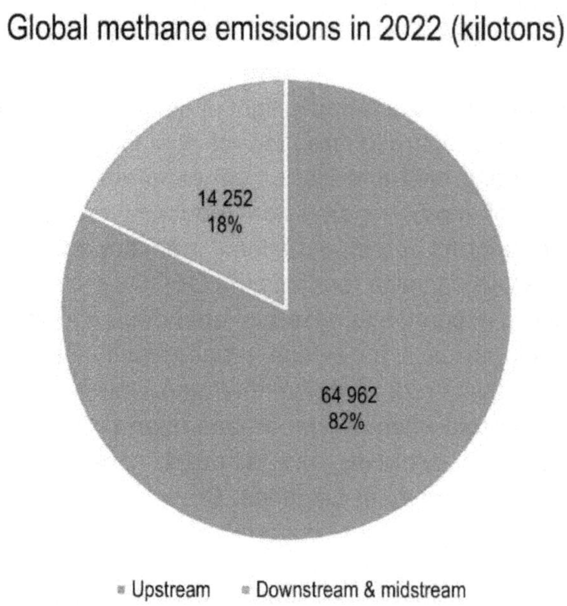

Source: Adapted from (IEA, 2023[4]).

Structure of the report and next steps

This report sets out recommendations to guide developing oil and gas producing countries through the regulatory actions they can take to reduce methane emissions in the upstream oil and gas sector. In that regard, this report sets out **two initial steps** followed by **five regulatory building blocks**. The two steps recommend the development of methane emission inventories and baselines and the setting of specific methane reduction targets embedded in low-emissions development trajectories and nationally determined contributions (NDCs) to pursue an integrated climate, energy and development agenda. The building blocks identify the key components for the design of robust upstream methane emission regulations. This includes guidance on how to set measurement/monitoring, reporting, and verification requirements; setting planning and licencing requirements; setting requirements to reduce flaring and venting; and addressing fugitive methane emissions. In recognition that purely prescriptive approaches may not be effective in developing country contexts to deliver deep cuts to methane emissions, the report offers guidance on a recommended reward-penalty system, providing incentives for methane abatement to effectively drive collective action on methane mitigation. Lastly, recognising that the rolling out of methane emission reduction technology and practices across the upstream oil and gas sector come with associated costs, the report **maps out different sources of finance** that developing oil and gas producing countries may be able to draw from.

Each section of the report builds on the recommendations of Pillar 1 of the *Equitable Framework and Finance for Extractive-based Countries in Transition (EFFECT)* on upstream oil and gas decarbonisation and offers country practice examples, drawing on existing regulations in different oil and gas jurisdictions (OECD, 2022[17]). The analysis covers regulations from both developed and developing economies and regulations administered at both the national and sub-national level. It also draws on the current tools, databases and research, including the IEA's *Methane Regulatory Roadmap and Toolkit, Methane policy and regulation database*, and *Methane Tracker*, the Global Flaring and Methane Reduction Partnership (GFMR)'s *Global Flaring and Venting Regulations* and the (CCAC) *Resource Library*. Alongside concrete regulatory examples, this report also takes into account relevant methane industry initiatives and scientific data made available through existing projects such as the UNEP's IMEO.

This report can be used as a basis for knowledge sharing, mutual learning and multi-stakeholder dialogue on regulatory action on methane emissions reduction. Its objective is to support engagement with developing countries producing oil and gas to "accelerate and substantially reduce non-carbon-dioxide emissions globally, including in particular methane emissions by 2030", as agreed at COP28, while also preserving the competitiveness of their gas exports in light of emerging methane import requirements. This report further aims to support the operationalisation of the Global Methane Pledge, moving from commitments to action and emerging methane abatement exporter-importer cooperative frameworks by highlighting concrete examples of methane emission reduction requirements and incentives across jurisdictions to inform regulatory reforms as part of national methane action plans and sustainable energy pathways. The results of the analysis can also feed into the OECD's Inclusive Forum on Carbon Mitigation Approaches (IFCMA) by reviewing mitigation approaches on methane emission reduction in the upstream oil and gas sector and by providing a basis for evidence-based mutual learning and inclusive multilateral dialogue for all countries to collectively reduce global methane emissions at least 30% from 2020 levels by 2030. The report can also facilitate international co-operation by harmonising approaches wherever feasible, while considering specific contexts and different capabilities, as well as public-private collaboration for effective methane abatement solutions. Country platforms on methane abatement and just transition pathways can help co-ordinate government strategies, donors, development finance institutions, industry and private investors. Through these platforms, sustainable investment and financial flows can be aligned, while managing the provision of technical assistance and capacity-building to support developing countries producing oil and gas in their transition journey.

References

Balcombe, P. et al. (2015), "Methane and CO2 Emissions from the Natural Gas Supply Chain an evidence assessment", *Sustainable Gas Institute*, https://www.imperial.ac.uk/sustainable-gas-institute/our-research/white-paper-series/white-paper-1-methane-and-co2-emissions-from-the-natural-gas-supply-chain/. [26]

Balcombe, P., N. Brandon and A. Hawkes (2018), "Characterising the distribution of methane and carbon dioxide emissions from the natural gas supply chain", *Journal of Cleaner Production*, Vol. Volume 172, pp. 2019-2032, https://www.sciencedirect.com/science/article/pii/S0959652617328962. [25]

Cahill, B. and A. Dawes (2023), *Transatlantic Efforts to Cut Methane Emissions*, Center for Strategic and International Studies (CSIS), https://csis-website-prod.s3.amazonaws.com/s3fs-public/2023-06/230623_Cahill_Transatlantic_Emissions.pdf?VersionId=7IKbvBjW88GlhmQVZaReNVltWHUrtB6W. [1]

CLDP (2023), *Methane Abatement Handbook*, U.S. Department of Commerce, https://cldp.doc.gov/sites/default/files/2023-09/CLDP%20Methane%20Abatement%20Handbook.pdf. [23]

Connoy, J., J. McKenzie and J. Gorski (2022), *Success in Eliminating Methane in Alberta's Peace River Region*, Pembina Institute, https://www.pembina.org/pub/methane-peace-river. [18]

Di Odoardo, M. et al. (2024), *Call of duties: how emission taxes on imports could transform the global LNG market*, Wood MacKenzie, https://www.woodmac.com/horizons/emission-taxes-could-transform-global-lng-market/. [16]

European Commission (2023), *Commission welcomes deal on first-ever EU law to curb methane emissions in the EU and globally*, European Commission, https://ec.europa.eu/commission/presscorner/detail/en/IP_23_5776. [15]

Global Methane Pledge (2021), *Global Methane Pledge*, https://www.globalmethanepledge.org/. [9]

IEA (2024), *Gas Market Report, Q1-2024*, International Energy Agency, Paris, https://www.iea.org/reports/gas-market-report-q1-2024. [14]

IEA (2024), *Global Methane Tracker 2024*, International Energy Agency, Paris, https://www.iea.org/reports/global-methane-tracker-2024. [2]

IEA (2023), *Global Methane Tracker 2023*, International Energy Agency, Paris, https://www.iea.org/reports/global-methane-tracker-2023. [4]

IEA (2023), *The Imperative of Cutting Methane from Fossil Fuels*, International Energy Agency, Paris, https://www.iea.org/reports/the-imperative-of-cutting-methane-from-fossil-fuels. [7]

IEA (2021), *Net Zero by 2050*, International Energy Agency, Paris, https://www.iea.org/reports/net-zero-by-2050. [6]

IGSD (2023), *A Primer on Cutting Methane: The Best Strategy for Slowing Warming in the Decade to 2030*, Institute for Governance & Sustainable Development, https://www.igsd.org/wp-content/uploads/2022/09/IGSD-Methane-Primer_2022.pdf. [22]

Jeudy-Hugo, S. (2024), “Insights for designing mitigation elements in the next round of Nationally Determined Contributions (NDCs)”, *OECD/IEA Climate Change Expert Group Papers*, No. 2024/01, OECD Publishing, Paris, https://doi.org/10.1787/b70a88ef-en. [3]

Kayrros (2023), *Global Methane Pledge Signatories Failing In Bid To Cut Emissions By 30% By 2030, Kayrros Data Shows*, Kayrros, https://www.kayrros.com/global-methane-pledge-signatories-failing-in-bid-to-cut-emissions-by-30-by-2030-kayrros-data-shows/. [11]

OECD (2022), *Equitable Framework and Finance for Extractive-based Countries in Transition (EFFECT)*, OECD Development Policy Tools, OECD Publishing, Paris, https://doi.org/10.1787/7871c0ad-en. [17]

Olczak, M., A. Piebalgs and P. Balcombe (2023), *A global review of methane policies reveals that only 13% of emissions are covered with unclear effectiveness*, One Earth, https://www.sciencedirect.com/science/article/pii/S2590332223001951. [12]

Ravikumar, A. (2020), *Repeated leak detection and repair surveys reduce methane emissions over scale of years*, Environmental Research Letters, https://iopscience.iop.org/article/10.1088/1748-9326/ab6ae1. [20]

Reuters (2023), *New Mexico’s rules help curb methane emissions more than Texas- Kayrros*, Reuters, https://www.reuters.com/world/us/new-mexicos-rules-help-curb-methane-emissions-more-than-texas-kayrros-2023-11-08/. [19]

SEI, Climate Analytics, E3G, IISD, UNEP (2023), *The Production Gap: Phasing down or phasing up? Top fossil fuel producers plan even more extraction despite climate promises*, Stockholm Environment Institute, Climate Analytics, E3G, International Institute for Sustainable Development, United Nations Environment Programme, https://doi.org/10.51414/sei2023.050. [5]

The White House (2016), *Leaders’ Statement on a North American Climate, Clean Energy, and Environment Partnership*, Office of the Press Secretary, The White House, https://obamawhitehouse.archives.gov/the-press-office/2016/06/29/leaders-statement-north-american-climate-clean-energy-and-environment. [8]

Tran, T. et al. (2022), “Leak detection and repair data from California’s oil and gas methane regulation show decrease in leaks over two years”, *Environmental Challenges*, Vol. 8, https://doi.org/10.1016/j.envc.2022.100563. [21]

U.S. Department of Energy (2023), *Greenhouse Gas Supply Chain MMRV Framework*, https://www.energy.gov/sites/default/files/2023-09/MMRV%20Framework%20US%20Stakeholder%20Meeting_Final_15Sept2023.pdf. [24]

UAE COP Presidency (2023), *Oil & Gas Decarbonization Charter launched to accelerate climate action*, UAE COP Presidency, https://www.cop28.com/en/news/2023/12/Oil-Gas-Decarbonization-Charter-launched-to--accelerate-climate-action. [10]

UNEP (2023), *An Eye on Methane — The road to radical transparency: International Methane Emissions Observatory*, United Nations Environment Programme, https://www.unep.org/resources/report/eye-methane-international-methane-emissions-observatory-2023-report. [13]

2 Step 1: Building emissions national inventories and baselines

Box 2.1. EFFECT recommendations: National inventories and baselines

What can governments do?

- Develop an emissions profile to identify how much methane is emitted and determine the location of the biggest sources, measure to the extent possible and estimate the level of emissions
- Develop protocols for incorporating new data such as satellite, flyovers and on-the-ground surveys, into national inventories
- Regulators should consider new technologies such as continuous monitoring systems, aerial surveillance, and satellite instruments as independent sources of data
- Collect information on flaring and venting, by requiring oil and gas companies, including NOCs, to publicly disclose such information
- What can governments and the fossil fuel industry do together?
- Establish a collaborative process to improve national inventories reports for oil and gas methane emissions by defining the different categories of emissions, reviewing the approach to emission estimation and data compilation, and updating the process after construction of the inventory.

Source: (OECD, 2022[1]).

Developing an emissions profile

At the outset, it is critical for governments to understand the nature and magnitude of methane emissions to determine the location of the biggest sources, to measure to the extent possible and estimate the level of emissions in order to design a robust regulatory framework that responds to the specific characteristics of the upstream oil and gas sector in their jurisdiction and better target regulatory intervention. For example, Canada's 2018 methane regulations adopts a "potential to emit" threshold which exempts facilities producing less than 60 000 m^3 of gas per year from certain regulatory requirements (venting limits, LDAR and equipment standards) but can achieve higher total methane reductions, rather than applying the same rules uniformly across all the emitting sources (Olczak, Piebalgs and Balcombe, 2023[2]).

The measurement and reporting of methane emissions is also an important component of the *EU Regulation on methane emissions reduction in the energy sector* which imposes specific requirements on oil and natural gas that is imported to Europe. Oil and gas producing developing countries that intend to export oil and gas to the European Union should ensure that their regulatory framework for measuring, monitoring and reporting methane emissions is equivalent to Article 12 of the EU's regulations or meets the requirements of OGMP 2.0 level 5 – for the requirements of OGMP 2.0 reporting levels, see Table 2.1. Article 12 of the EU's regulations also provides that monitoring and reporting requirements ensure at least source and site level quantification, as well as regular reporting. In addition, producer governments will need to implement either a third-party verification scheme – where independent accredited verifiers review emissions reports prepared by operators to assess their accuracy and credibility; or a government verification and auditing scheme – where the government empowers a relevant regulatory agency (e.g. an environmental protection agency) to undertake verification and auditing activities.

Table 2.1. OGMP 2.0 Reporting levels

LEVEL 1	LEVEL 2	LEVEL 3	LEVEL 4	LEVEL 5
Venture/Asset Reporting	**Emissions Category**	**Generic Emission Source Level**	**Specific Emission Source Level**	**Level 4 + Site Level Measurement Reconciliation**
Single, consolidated emissions number	Emissions reported based on IOGP and Marcogaz emissions categories	Emissions reported by detailed source type	Emissions reported by detailed source type using specific emissions and activity factors	Level 5: Integrating bottom-up source-level reporting (L4) with independent site-level measurements.
Only applicable where company has very limited information	Based on generic emissions factors	Based on generic emissions factors	Based on direct measurement or other methodologies	UNEP recommends attempts at site-level measurements with possible reconciliation for a nominal 1/3 of assets or emissions with subsequent year-over-year progress to move all material assets to L5. Site-level measurements: direct measurement technologies at a site or facility level on a representative sample of facilities

Source: (UNEP, 2020[3]).

Historically, methane emissions measurements were based on unquantified estimations or emission factor-based calculations (multiplying activities by emission factors), often resulting in incomplete or inaccurate information. However, in recent years with advancements in technology and increasing international co-operation, governments have access to new sources of information that can be used to improve the accuracy of methane detection and monitoring – see Box 2.2.

Box 2.2. Monitoring methane emissions by satellite

Satellites are a key tool in monitoring methane emissions and are becoming increasingly accessible to governments and companies by providing better and more transparent data to inform methane mitigation measures. However, while satellites can complement existing ground-based measurement practices, they may provide insufficient data in some situations. For example, satellites can struggle to provide readings in some environments including offshore areas, mountainous regions or at high latitudes. Satellite readings may also be negatively affected by cloud cover, and this is particularly relevant in areas with dense forests or those in equatorial regions, such as Nigeria or Venezuela.

Developing countries can benefit from the increasing availability of methane detection satellites and associated detection programmes to monitor methane emissions in their jurisdictions. Some of the more prevalent methane detection satellites and providers include:

- **GHGSat** – a satellite-based and aerial remote sensing company with expertise in monitoring methane emissions. In 2023, GHGSat's satellites carried out around 13 000 daily observations at specific oil, gas and coal facilities across Eurasia, North America, the Middle East and Australia. In 2021, GHGSat partnered with OGCI to detect and characterise previously unknown persistent methane emissions sources. Operators provided on the ground confirmation for several persistent large sources detected through satellite monitoring. Data is free and available upon request.
- **Methane Alert and Response System (MARS**) – UNEP's MARS programme was launched at COP27 to facilitate the use of satellites to detect major methane emission events and to then alert government authorities and relevant operators. In 2023, MARS detected 500 large methane emissions events from global oil and gas operations and delivered approximately 80 notifications to relevant countries and operators. Data is free and publicly available.
- **MethaneSAT –** backed by the Environmental Defense Fund (EDF), MethaneSat was launched in March 2024 to provide frequent and high spatial resolution coverage of methane emissions from oil and gas facilities. MethaneSat is designed to regularly monitor roughly 50 major regions accounting for more than 80% of global oil and gas production. Data is free and publicly available.
- **Sentinel programme** – the European Space Agency's Copernicus programme aims to achieve a global, continuous, autonomous and high-quality Earth observation capacity. Central to this programme are the Sentinel satellites which provide optical imaging for land services, and monitoring of weather and atmospheric conditions. Several Sentinel satellites, in particular, Sentinel-2 and Sentinel-3 and Sentinel-5P, provide detection/monitoring of methane emissions, and are used by several public and private actors including the IMEO and analytics firm Kayrros. For example, in 2023, using data from Sentinel-5P, Kayrros released a tool that quantifies large methane emissions and provides country-level oil and gas methane intensities. Kayrros detected 152 methane super-emitter events in the United States from oil and gas operations. Data is free and publicly available.

Source: (IEA, 2024[4]); (OGCI, 2024[5]).

Governments can use two approaches to measure methane emissions:

- **National inventory approach (bottom-up)** – this approach is already used by countries to submit their emission inventories to the UNFCCC. Under the UNFCCC process, governments are expected to track and report national-level GHG emissions through the development of inventories. These inventories cover methane as well and can be used as a basis for building a specific methane emissions inventory. While the Intergovernmental Panel on Climate Change (IPCC) has issued guidelines to ensure that methane emission factors account for local characteristics (e.g. surface versus underground mine) (Vernon et al., 2022[6]), these inventories normally use estimates via bottom-up approaches where a typical emissions factor is applied to the total number of oil and components (e.g. wellheads, pneumatic devices, storage tanks) to estimate total emissions, rather than direct measurements;

- **Atmospheric observations approach (top-down)** – this approach uses remote sensing from towers, aerial surveys and satellites to monitor emissions of individual facilities and regions. This approach has been used to infer historical global emission trends, and can lead to better confidence in the outcomes of specific emissions reduction efforts (Vernon et al., 2022[6]; UNEP, 2022[7]; IEA, 2021[8]).

Recent studies in the oil and gas sector have identified discrepancies between inventories compiled with bottom-up and top-down emission methods, based directly on empirical, atmospheric data of emitting units or facilities (e.g. aerial surveys and satellites). For example, satellite-based estimates of methane emissions from the oil and gas sector in Mexico are 100% higher than emissions presented in the Mexican national greenhouse gas inventory (UNEP, 2022[7]; CLDP, 2023[9]). One possible explanation for such discrepancies is that most inventories generated by countries and companies rely on bottom-up approaches that are not based on recent measured data. These inventories may be underestimating potential emissions sources, and in particular may be missing “super-emitting” events which are often responsible for an outsized share of overall emissions levels (IEA, 2024[4]).

Although bottom-up inventories often underestimate overall emissions, they can still provide a useful starting point for methane emissions measurement. The creation of a granular asset and equipment level inventory can be an important first step and can encourage collaboration between the NOC and/or energy ministry, who may hold some of this data, and the environment/climate change ministry that may not. Furthermore, the creation of a bottom-up inventory can help the country to identify and prioritise next steps – for example, where regulatory action can proceed without having to wait for top-down emissions data. The results of this bottom-up equipment inventory, even with generic emission factors, can also help the country understand where to focus measurement studies to improve upon generic emission factors. Finally, the top-down site level measurement technologies can help validate the results of the bottom-up inventory, highlight areas of discrepancy needing more research, and potentially track reductions associated with regulatory compliance. Governments can make use of publicly available tools to guide them through the process of setting up a bottom-up equipment inventory – see Box 2.3.

Box 2.3. CATF's Country Methane Abatement Tool

The Country Methane Abatement Tool (CoMAT) is a tool designed to make it easier for countries to quickly estimate their methane emissions and abatement potential, develop comprehensive mitigation approaches, and design methane reduction policy strategies.

Launched in 2019, CoMAT lets users develop initial and refined estimates of their jurisdictions' emissions and reduction potential using the best information the users have available and provides access to a digitised library of leading methane policy and proven best practices.

CoMAT's emissions calculations and inventory tools can help countries understand what information is needed to put together a bottom-up inventory of emissions. Furthermore, CoMAT can assist countries to continually refine their emissions inventories and explore variables and specific policy and regulatory options that can drive pollution reduction.

Source: (CATF, 2019[10]).

Ensuring regulations can accommodate new technologies to supplement national inventories

Governments can rely on a number of traditional and new technologies to inform the development of national inventories. In recent years a number of technologies that can detect and measure methane emissions have been developed, including sensors that can be mounted on satellites, aircraft, drones, or vehicles, as well as sensors that can be permanently installed at a single site/facility to provide more near-continuous monitoring of methane emissions.

Governments should ensure that regulations are crafted to accommodate the use of new technologies. This could be done through the inclusion of emissions reduction performance standards, as opposed to prescribing the deployment of specific technologies. In this regard, governments should monitor developments in methane measurement technologies to ensure that regulatory requirements do not inadvertently lock in old technologies and prevent uptake of new options (IEA, 2021[8]). For example, no-fly zones around production sites or other oil and gas infrastructure may limit drones and non-government use of satellite imagery from being used to detect methane emissions. Governments may also need to ensure that regulators are empowered and resourced to engage with third-party technology providers to conduct aerial or satellite monitoring, and to engage with international organisations and non-governmental organisations that provide free aerial surveys and satellite monitoring of methane emissions – for example, the OGCI's Satellite Monitoring Campaign (SMC) and the CCAC Methane Science Program (CLDP, 2023[9]). Regulators may also need to build capacity to interpret the results and make effective use of the data.

Setting corporate disclosure requirements to inform emissions baselines and national inventories

Governments should work with the oil and gas industry to establish a collaborative process to improve national inventories reports for oil and gas methane emissions. This process should include defining the different categories of emissions, reviewing the approach to emission estimation and data compilation, and updating the process after construction of the inventory.

Box 2.4. Addressing super-emitter events through regulation

The U.S. Environmental Protection Agency (EPA) developed the super-emitter programme in response to studies indicating that large, irregular emissions events contribute almost 50% of methane emissions from the oil and gas sector.

As set out in the U.S. EPA's 2023 *Final Rule to Reduce Methane and Other Harmful Pollution from Oil and Natural Gas Operations,* that entered into force on 7 May 2024, the super-emitter programme serves as a backstop to other reporting provisions of the Final Rule in that it allows EPA-certified third parties to supplement a facility's required routine monitoring using EPA approved, remote, advanced sensing technologies capable of identifying an on-going super emissions event.

The approved third parties must notify EPA within 15 days of discovering an ongoing event and the EPA will then review the data for completeness and accuracy "to a reasonable degree of certainty". If the data meet EPA's criteria, EPA will notify the owners or operator of the event who must initiate an investigation with five days of notification.

Source: (Jenks, Dobie and Leahy, 2023[11]).

While governments have the responsibility of setting regulations, operators are likely to have better information than governments about the nature and extent of their methane emissions profile. In particular, operators may have information about the distribution of emissions at the site/facility-level – which is key in order to identify high-emitting facilities (super emitters) that provide a disproportionate contribution to total emissions in the jurisdiction (UNEP, 2022[7]).

Examples of collaborative public-private efforts to measure emissions can be seen in methane regulations in Colombia and Mexico that set out requirements for the development of an emissions profile. These include specific requirements for the establishment of a methane emissions baseline to identify, classify, and quantify methane emissions, and to serve as a reference for the comparison of methane emissions reductions in the subsequent years. For example, in Mexico operators must establish a baseline of natural gas emissions for each facility that includes a diagnosis of emissions that occur in equipment, components, and wells operations. The baseline should include: the identification, classification, and quantification of methane emissions. The emissions baseline must be reported to the regulator and will serve as a reference for the comparison and for the continuous improvement of methane emissions reductions for the subsequent years.

Similarly, Colombia's 2018 methane regulations specify that an operator must establish a baseline of natural gas emissions for each facility and that includes all equipment and components. Operators must submit an emissions baseline to the regulator within 30 days from preparation for approval. The baseline will serve as a reference for the comparison and for the continuous improvement of methane emissions reductions for the subsequent years, as baselines must be updated every three years and re-submitted for approval. The baseline will cover emissions from: oil and gas production tests, well completions, discharge of liquids in exploration and production wells, well pilot testing, well stimulation including hydraulic stimulation and return fluid injection, well service, well abandonment and well workover activities. Article 45 provides that operators must establish a baseline within 12 months from the start of operations for new facilities and within 24 months after the entry into force of the regulation for existing facilities.

References

CATF (2019), *Country Methane Abatement Tool*, Clean Air Task Force, https://www.catf.us/comat/. [10]

CLDP (2023), *Methane Abatement Handbook*, U.S. Department of Commerce, https://cldp.doc.gov/sites/default/files/2023-09/CLDP%20Methane%20Abatement%20Handbook.pdf. [9]

IEA (2024), *Global Methane Tracker 2024*, International Energy Agency, Paris, https://www.iea.org/reports/global-methane-tracker-2024. [4]

IEA (2021), *Driving Down Methane Leaks from the Oil and Gas Industry*, International Energy Agency, Paris, https://www.iea.org/reports/driving-down-methane-leaks-from-the-oil-and-gas-industry. [8]

Jenks, C., H. Dobie and R. Leahy (2023), *EPA's Final Methane Rule—Incorporating Advanced Technologies and Emissions Data to Reduce Methane Emissions from the Oil and Natural Gas Sector*, Environmental and Energy Law Program, Harvard Law School, https://eelp.law.harvard.edu/wp-content/uploads/2024/08/EELP-EPA-Final-Methane-Rule.pdf. [11]

OECD (2022), *Equitable Framework and Finance for Extractive-based Countries in Transition (EFFECT)*, OECD Development Policy Tools, OECD Publishing, Paris, https://doi.org/10.1787/7871c0ad-en. [1]

OGCI (2024), *White Paper: Results of OGCI Satellite Monitoring Campaign 2022-2023 over Kazakhstan, Algeria and Egypt*, Oil and Gas Climate Initiative (OGCI), https://www.ogci.com/wp-content/uploads/2024/03/SMC-Results-v4.pdf. [5]

Olczak, M., A. Piebalgs and P. Balcombe (2023), *A global review of methane policies reveals that only 13% of emissions are covered with unclear effectiveness*, One Earth, https://www.sciencedirect.com/science/article/pii/S2590332223001951. [2]

UNEP (2022), *An Eye on Methane: International Methane Emissions Observatory*, United Nations Environment Programme, https://www.unep.org/resources/report/eye-methane-international-methane-emissions-observatory-2022-report. [7]

UNEP (2020), *Mineral Methane Initiative: OGMP 2.0 Framework*, United Nations Environment Programme, https://ogmpartnership.com/wp-content/uploads/2023/02/OGMP_20_Reporting_Framework-1.pdf. [3]

Vernon, N. et al. (2022), *How to Cut Methane Emissions*, International Monetary Fund, https://www.elibrary.imf.org/view/journals/066/2022/008/066.2022.issue-008-en.xml. [6]

3 Step 2: Setting national sector-specific methane reduction targets

Box 3.1. EFFECT recommendations: Methane reduction targets

What can governments do?

- Integrate methane emissions reduction into NDC targets and implementation plans.
- Set progressively ambitious methane emissions reduction targets.
- Require the public disclosure of methane emissions information, including publication on a website. This added layer of scrutiny may create an additional incentive for companies to comply.
- Publicly disclose progress made toward methane emissions reduction targets on a regular basis.

Source: Adapted from (OECD, 2022[1]).

Governments can send a strong signal of their intention to tackle methane emissions across the upstream oil and gas sector by setting out national sector-specific methane emission reduction targets, and by explaining how these targets contribute, in a nationally determined manner, to global mitigation efforts aligned with 1.5°C pathways, consistent with the Paris Agreement first global stocktake outcome adopted at COP28, calling for accelerating the substantial reduction of non-carbon-dioxide emissions globally, in particular methane emissions by 2030, while also transitioning away from fossil fuels in energy systems, in a just, orderly and equitable manner.

Targets may be voluntary or binding and can be expressed as tonnes of methane reduced, a percentage reduction below historic emission levels or a declining intensity ratio (i.e. methane emissions per unit of production).

Targets for methane (or GHG) emissions can be defined as absolute targets or as intensity target:

- ***An absolute target*** refers to a target that aims to reduce emissions by a set amount.
- ***Intensity-based targets*** are expressed as units of emissions per unit of activity. Activity can be measured at an aggregate level, for example in terms of GDP or per capita GDP, or at a more detailed level based on measures of underlying efficiency of the economy.

Intensity-based targets do not necessarily cap emissions. In fact, even if intensity targets are achieved, emissions may grow as the economy grows. For intensity-based targets to deliver absolute emission reductions the targets should be demanding enough: if the rate of decline in emissions intensity is higher than the rate of GDP growth, then absolute emissions will fall. An alternative approach is to combine intensity targets with absolute targets or caps (OECD, 2022[1]).

A number of voluntary international initiatives have emerged over the past few years to encourage the uptake of methane emissions targets. In 2019, the multilateral Global Methane Alliance (GMA) was launched to bring together governments, financing institutions, international organisations and NGOs, and industry to support ambitious methane reduction targets from the oil and gas industry and has called on countries to set targets of at least 45% emission reduction from 2005 levels by 2025, and 60-75% by 2030.

Setting methane emissions reduction targets are an important signal of governments' intention to tackle methane emissions across the upstream oil and gas sector. However, in order to demonstrate further commitment, governments should also track and make public the status of progress toward these methane emissions reduction targets. This can add an additional layer of accountability and transparency as well as provide input for setting future progressively ambitious targets.

Alongside governments' disclosure of methane emissions reduction targets, regulations can also require oil companies to publicly disclose their methane targets as well as progress toward those targets in that jurisdiction. Many IOCs have already publicly committed to various methane targets – either individually or as part of membership in an industry organisation or adherence to an international standard. For example, the Oil & Gas Decarbonization Charter (OGDC) was launched at COP28 where participants agreed to aim for near-zero methane emissions by 2030 at operated facilities and to engage with joint operating partners to achieve near-zero methane emission at non-operated facilities. The Charter's signatories include 31 NOCs and 22 IOCs – collectively representing more than 42% of global oil production (OGDC, 2023[2]).

In addition, the OGCI, whose member companies includes 12 large IOCs and NOCs, launched the Aiming for Zero Methane Emissions initiative in 2022, setting a 2025 methane intensity target of well below 0.20% from 2017 levels with a view to reach near zero methane emissions from operated assets by 2030 (OGCI, 2022[3]). OGCI methane intensity KPI is the sum of methane emissions of all upstream operated emissions of OGCI companies divided by the volume of natural gas sold. The denominator excludes oil volumes, so this KPI does not apply to pure oil players or specific assets that produce oil. OGCI member companies collectively report within OCGI their methane emissions both on an intensity basis and on an absolute basis. From 2017-2022, OGCI found that member companies halved emissions on both an absolute and on an intensity basis, with methane intensity ratios declining from 0.3% to 0.15%. However, it should be noted that this target was met collectively by OGCI as a group. Some OGCI companies may not have met target as individual OGCI companies, and will continue their efforts to reduce methane emissions.

However, the lack of sufficient enabling policies and regulations that incentivise company decarbonisation may be factors that could prevent some of the companies operating in those jurisdictions from being able to comply with a 1.5°C trajectory.

Setting methane reduction targets in Nationally Determined Contributions

At COP28, countries agreed to submit by February 2025 updated nationally determined contributions (NDCs 3.0), that are economy-wide, cover all GHGs, and are aligned with limiting global warming to 1.5°C to drive action on emissions reduction until 2035. Current NDCs present significant gaps in coverage from a global emissions perspective. For example, 15% of Parties, accounting for 46% of total global methane emissions in 2020, communicate corresponding measures for reducing methane emissions from fossil fuel operations (UNFCCC, 2023[4]). The first global stocktake recognised the urgent need to address persisting gaps to deliver the course correction needed to keep the 1.5°C goal within reach. The next round of NDCs is a significant opportunity for governments to set ambitious targets on methane mitigation in the oil and gas sector to address persisting gaps and lay out plans to achieve them, including means of implementation such as investment and finance.

The first global stocktake further encourages all Parties to align their NDCs with their Long-term Low GHG Emission Development Strategies (LT-LEDS). Embedding LT-LEDS implies articulating how natural gas production and use contribute to the achievement of climate objectives goals, sustainable development priorities and energy security needs. Anchoring methane abatement projects within the oil and gas sector decarbonisation pathways and just transition strategies will send clear signals to investors, help mobilise financial support, and contribute to broader systemic transformation. Sectoral analysis underpinning the preparation of NDCs provides the opportunity to explain how the oil and gas sector contributes to the overall economy-wide emission reduction targets, underpinned by robust implementation and investment plans that can support the delivery of national climate commitments.

For example, countries may adopt national methane action plans[1] to clearly set out economy-wide ambitions for controlling methane emissions. National methane action plans often include high-level reduction targets for specific sectors and may also list specific mitigation measures that will be enacted in pursuit of these targets. For example, Canada's *Methane Strategy 2022* includes a commitment to reduce oil and gas sector emissions by 75% by 2030 compared to 2012 levels, alongside specific abatement measures and supporting programmes. In Viet Nam, the *Methane Action Plan 2030* sets out targets to reduce economy-wide methane emissions by at least 30% below 2020 levels by 2030, with specific targets for the agriculture, waste, and energy sectors. In recent years there has been a significant uptake in countries developing national methane action plans, driven in part by the increasing membership in the GMP. In May 2023, around 50 countries had either adopted or were developing national methane action plans (CLDP, 2023[5]). According to the IEA, if all methane pledges made by countries and companies so far are implemented in full and on time, methane emissions from oil and gas would reduce by 55% by 2030. However, at present, there remains a gap between ambition and implementation as only one-third of the 156 members of the GMP have outlined sector-specific methane targets or have developed national methane action plans (IEA, 2024[6]).

Box 3.2. National methane action plans

Côte d'Ivoire's National Action Plan on Short-lived Climate Pollutants

In 2019, Côte d'Ivoire released a *National Action Plan on Short-lived Climate Pollutants (NAP)* in order to align its short-term development objectives with its long-term aspirations to reduce greenhouse gas emissions. The objective of this NAP was to develop an inventory of emissions of short-lived climate pollutants and to identify measures to reduce emissions of these pollutants at the national level.

Côte d'Ivoire's NAP sets out specific measures to address methane emission from the oil and gas sector. This includes a programme to strengthen regulatory frameworks and technical capacity on fugitive emissions from venting, as well as improved control of unintentional leakages. In this regard, the NAP sets out the following goals for the oil and gas sector:

- Reduce 50% of avoidable fugitive emissions by 2030; and
- Reduce 70% of avoidable fugitive emissions by 2040.

Ghana's National Action Plan to Mitigate Short-Lived Climate Pollutants

In 2018, Ghana released a NAP to mitigate Short-Lived Climate Pollutants (SLCPs). The NAP exercise led to the identification and prioritisation of 16 short-lived climate pollutants mitigation measures across five main sectors: energy, transport, industrial process, agriculture, forestry and waste. The NAP includes an inventory of SLCPs across the economy and identifies methane as the most dominant SLCP as well as the second most important greenhouse gas after carbon dioxide.

Ghana's NAP notes the role of gas production on the release of SLCPs. Gas flaring is identified as a SCLP source activity, and the NAP recommends the implementation of a commissioning plan for the production of additional non-associated gas from the Sankofa fields.

Nigeria's National SLCP Action Plan

In 2019, Nigeria released a comprehensive National SLCP Action Plan (NAP) with the aim of reducing SCLPs and methane emissions by 61% by 2030. The NAP identifies 22 mitigation measures targeted at 8 different source sectors in order to reduce emissions from major SLCPs including methane, as well as reducing emissions of co-emitted long-lived greenhouse gases such as carbon-dioxide and other air pollutants.

Nigeria's NAP sets out specific abatement measures to address methane emissions from the oil and gas sector. These include the elimination of gas flaring, control of fugitive emissions from oil production and processing, and reductions in methane leakage from transportation and distribution. The NAP sets out targets for each of the 22 abatement measures.

Source: (MINEDD, 2019[7]); (EPA, 2018[8]); (Federal Ministry of Environment, 2019[9]).

Incorporating methane reduction targets in NDCs, can be a useful tool to align gas development with climate objectives. NDCs not only reflect a country's ambitions for mitigation but also take into account its domestic context and capabilities. As such, they may be expressed as conditional or unconditional. Unconditional targets refer to measures and actions that a country can implement based on its own capabilities. Whereas a conditional NDC, refers to areas where international support is required in order for a country to meet that commitment.

Table 3.1. GHG and methane emissions reduction targets in NDCs

	Reduction target	Date	Nature of commitment	NDC submission date
Algeria	Reduce gas flaring to 1%	2030	Unconditional	20 October 2016
Angola	Reduce flaring – 490 MMSCF/day	2030	Unconditional	31 May 2021
	Reduce flaring – 110 MMSCF/day	2030	Conditional	
Brunei	Zero routine flaring	2030	Unconditional	31 December 2020
Egypt	65% reduction in GHG emissions (from 2015 baseline)	2030	Conditional	26 June 2023
Gabon	63% reduction in GHG emissions from flaring (from 2000 baseline)	2025	Unconditional	2 November 2016
Ghana	20% reduction in fugitive methane from oil and gas infrastructure (from 2019 baseline)	2030	Unconditional	4 November 2021
Mexico	Reduce GHG emissions from the oil and gas sector by 14%	2030	Unconditional	17 November 2022
Nigeria	Zero gas flaring	2030	Conditional	30 July 2021
	60% reduction in fugitive methane emissions	2031	Conditional	
Oman	Zero routine flaring	2030	Unconditional	29 July 2021
	Reduce GHG emissions from the oil and gas sector by 17% (from 2021 baseline)	2030	Mixed	29 November 2023
Qatar	Zero routine flaring	2030	Unconditional	24 August 2021
Saudi Arabia	Zero flaring	2030	Unconditional	23 October 2021

Note: Qatar's NDC commitment is made in respect of Qatar's NOC (Qatar Energy).
Source: (UNFCCC, 2023[10]); (World Bank, 2022[11]).

As of 2024, there are more than 190 NDCs in place. These include around 35 with specific targets for reducing methane emissions (from all sectors of the economy) and 20 which set out specific measures to reduce methane emissions from fossil fuels (IEA, 2024[6]). While many oil and gas producing countries have committed to reducing methane emissions in their NDCs,[2] only a few have set specific oil and gas sector methane reduction targets in their NDCs (World Bank, 2022[12]). In some cases, countries have adopted broader GHG reduction targets which can include methane reduction commitments indirectly – for example commitments to zero routine flaring are designed to prevent the release of CO_2 but will cover methane as well as small amounts of methane are emitted during flaring unless the flare efficiency is 100%. Ghana and Nigeria are the only countries identified that include specific methane reduction targets in their NDCs – see Table 3.1 above.

The adoption of methane emission reduction targets provides clear policy direction for the development of regulations to achieve set emission reduction targets – see Box 3.3 for the approach taken in Nigeria.

Box 3.3. Nigeria: From methane reduction targets to comprehensive upstream regulations

Nigeria is the world's thirteenth-largest oil and gas producer, with an average production of 1 268 000 barrels of crude oil per day in 2023. Natural gas production takes place on smaller scale, with Nigeria producing 45.9 bcm of natural gas in 2021. Nigeria has consistently been among the world's top seven largest gas flaring countries by volume and emitted 3 306 kt of methane from its energy sector in 2022.

Beginning in 2019, the Nigerian government, including the Federal Ministry of Environment and Nigeria's oil and gas regulator, collaborated with the Clean Air Task Force (CATF) to enhance Nigeria's national inventory for methane emissions (using CoMAT to calculate emissions) and to lay the groundwork for comprehensive policy development. On the back of these efforts:

- in 2021, Nigeria updated its NDC, conditionally committing to reducing fugitive methane emissions by 60% by 2031 and achieving net-zero by 2060 and joined the Global Methane Pledge at COP26; and
- in 2022, Nigeria became the first country in Africa to regulate methane emissions from its oil and gas sector, when the Nigerian Upstream Petroleum Regulatory Commission (NUPRC) released the *Guidelines for Management of Fugitive Methane and Greenhouse Gases Emissions in the Upstream Oil and Gas Operations in Nigeria*. The purpose of these guidelines is to achieve Nigeria's emission mitigation and reduction targets of the NDCs in the oil and gas sector. In this regard, the guidelines set out key abatement measures to eliminate routine gas flaring (100% by 2030) and control fugitive methane emissions (60% methane reduction by 2030). The guidelines also direct operators to prevent and control methane emissions through GHG management plans, monitoring and inspection requirements, and operational and equipment standards.

Source: (Trading Economics, 2023[13]); (Aizarani, 2023[14]); (World Bank, 2023[15]); (NUPRC, 2022[16]), (IEA, 2023[17]).

Setting progressive targets to reduce methane emissions

Developing producer countries should set specific and progressively ambitious methane emissions reduction targets. For example, in its first NDC submission in 2021, Angola committed to reduce flaring by 295 MMSCF per day over the period 2015-25 (as measured from its 2015 baseline), with a further reduction of 490 MMSCF per day by 2030. Nigeria's 2021 NDC conditionally commits to zero gas flaring by 2030 and to a 60% reduction of fugitive methane emissions by 2031 (UNFCCC, 2023[11]).

In the United Kingdom, progressive methane emissions reduction targets are set out in the North Sea Transition Deal – a plan for how the UK government and the offshore oil and gas sector will work together to deliver on GHG emissions reduction targets and to develop the required skills and infrastructure to meet these targets. The North Sea Transition Deal sets out GHG emissions reduction targets of 10% in 2025, 25% in 2027, and 50% in 2030 (as measured from its 2018 baseline) (BEIS & OGUK, 2021[18]).

Notes

[1] National methane action plans are often included within broader action plans to address other short-lived climate pollutants (SLCPs).

[2] For example, in Africa, all countries have committed to reducing methane emissions in their NDCs except Libya, South Sudan and Somalia (AfriCatalyst, 2023[19]).

References

AfriCatalyst (2023), *Mobilizing Finance for Methane Action in Africa*, AfriCatalyst, https://africatalyst.com/wp-content/uploads/2024/01/Mobilizing-Finance-for-Methane-Action-in-Africa.pdf. [19]

Aizarani, J. (2023), *Natural gas production worldwide in 2021, by country*, Statista, https://www.statista.com/statistics/264101/world-natural-gas-production-by-country/. [14]

BEIS & OGUK (2021), *North Sea Transition Deal*, Department for Business, Energy & Industrial Strategy, https://assets.publishing.service.gov.uk/media/605b148ce90e0724c7d30c2b/north-sea-transition-deal_A_FINAL.pdf. [18]

CLDP (2023), *Methane Abatement Handbook*, U.S. Department of Commerce, https://cldp.doc.gov/sites/default/files/2023-09/CLDP%20Methane%20Abatement%20Handbook.pdf. [5]

EPA (2018), *National Action Plan to Mitigate Short-Lived Climate Pollutants (Ghana)*, Environmental Protection Agency (EPA), https://www.ccacoalition.org/policy-database/national-action-plan-mitigate-short-lived-climate-pollutants-ghana. [8]

Federal Ministry of Environment (2019), *National Action Plan to reduce Short-Lived Climate Pollutants (SLCPs)*, Nigerian Federal Ministry of Environment, https://climatechange.gov.ng/wp-content/uploads/2020/09/nigeria-s-national-action-plan-nap-to-reduce-short-lived-climate-pollutants-slcps-.pdf. [9]

IEA (2024), *Global Methane Tracker 2024*, International Energy Agency, Paris, https://www.iea.org/reports/global-methane-tracker-2024. [6]

IEA (2023), *Global Methane Tracker 2023*, International Energy Agency, Paris, https://www.iea.org/reports/global-methane-tracker-2023. [17]

MINEDD (2019), *Cote d'Ivoire National SLCP Action Plan*, Ministry of Environment and Sustainaible development (MINEDD), https://www.ccacoalition.org/policy-database/summary-cote-divoire-national-slcp-action-plan. [7]

NUPRC (2022), *Guidelines for Management of Fugitive Methane and Greenhouse Gases Emissions in the Upstream Oil and Gas Operations in Nigeria*, Nigerian Upstream Petroleum Regulatory Commission, https://www.nuprc.gov.ng/wp-content/uploads/2022/11/METHANE-GUIDELINES-FINAL-NOVEMBER-10-2022.pdf. [16]

OECD (2022), *Equitable Framework and Finance for Extractive-based Countries in Transition (EFFECT)*, OECD Development Policy Tools, OECD Publishing, Paris, https://doi.org/10.1787/7871c0ad-en. [1]

OGCI (2022), *Aiming for Zero Methane Emissions Initiative*, Oil and Gas Climate Initiative, https://aimingforzero.ogci.com/. [3]

OGDC (2023), *Oil & Gas Decarbonization Charter*, Oil & Gas Decarbonization Charter, https://www.ogdc.org/about/. [2]

Trading Economics (2023), *Crude Oil Production – Africa*, Trading Economics, https://tradingeconomics.com/country-list/crude-oil-production?continent=africa. [13]

UNFCCC (2023), *Nationally Determined Contributions Registry*, United Nations Framework Convention on Climate Change, https://unfccc.int/NDCREG. [10]

UNFCCC (2023), *Nationally determined contributions under the Paris Agreement - Synthesis report by the secretariat*, UNFCCC Secretariat, Bonn, https://unfccc.int/sites/default/files/resource/cma2023_12.pdf. [4]

World Bank (2023), *Methane emissions in energy sector (thousand metric tons of CO2 equivalent) - Nigeria*, World Bank Open Data, https://data.worldbank.org/indicator/EN.ATM.METH.EG.KT.CE?end=2019&locations=NG&start=1990&view=chart. [15]

World Bank (2022), *Financing Solutions to Reduce Natural Gas Flaring and Methane Emissions*, World Bank, http://hdl.handle.net/10986/37177. [11]

World Bank (2022), *Global Flaring and Venting Regulations: A Comparative Review of Policies*, World Bank, https://thedocs.worldbank.org/en/doc/fd5b55e045a373821f2e67d81e2c53b1-0400072022/global-flaring-and-venting-regulations-a-comparative-review-of-policies. [12]

4 Regulatory building block 1: Setting measurement/monitoring, reporting, and verification requirements

Box 4.1. EFFECT recommendations: Methane measurement/monitoring, reporting and verification requirements

What can governments do?

- Establish a regulatory framework for the measurement, disclosure, reporting and verification of […] methane emissions, using existing reporting templates
- Require the public disclosure of methane emissions information, including publication on a website
- Establish a comprehensive framework […] that includes methane emissions reporting, recordkeeping and disclosure, and third-party verification requirements
- Introduce requirements for LNG project operators to measure emissions during liquefaction and transport.

Source: (OECD, 2022[1]).

Methane measurement, monitoring, reporting, and verification (MMRV) requirements are critical in order for governments and companies to understand the extent of their emissions profile, to monitor progress toward emissions reduction targets, and to design robust methane regulations. In this respect, MMRV requirements are an information-based regulatory approach, where regulated entities must quantify their emissions, either by measuring or estimating, and then report that data to the regulator – see Box 4.2.

Box 4.2. Methane measurement/monitoring, reporting and verification

Measurement / monitoring encompasses systematic observation, identification and assessment of methane sources, including purposeful venting, unlit flares, releases due to emergency situations, and fugitive emissions.

Reporting provisions require companies to send information to the regulator. Reporting requirements support compliance follow-up and help understand whether progress is being made. They are particularly relevant for the establishment of emission baselines.

Verification (or auditing) is a process where independent organisations or professionals observe and report on the validity of the information provided by oil and gas operators. This may include the use of accredited third-party verifiers or where the relevant regulatory agency (e.g. the country's environmental protection agency) carries out the verification and auditing itself.

Source: Adapted from (IEA, 2021[2]) and (Mohlin et al., 2022[3]).

Developing a regulatory framework for the measurement/monitoring, reporting and verification of methane emissions

The design and implementation of a robust and effective regulatory framework for the measurement, reporting and verification of methane emissions may place a significant burden on the resources and capabilities of many oil and gas developing producing countries. Fragmentation of efforts with the development of different methodologies across jurisdictions would also be counterproductive, limiting the usability of data for improved transparency and accountability. To avoid such suboptimal outcomes, governments can collaborate to develop MRV frameworks or can incorporate existing reporting standards and templates into their regulatory frameworks, in particular when seeking to align reporting requirements with those from importing jurisdictions to facilitate consistency, comparability and usability of reported data.

In terms of existing reporting standards and templates, UNEP's Oil & Gas Methane Partnership 2.0 (OGMP 2.0) represents the only comprehensive measurement-based reporting framework covering all material sources of methane emissions, from both operated and non-operated assets across all segments of the value chain. OGMP 2.0 sets out a framework for its member companies to report annually on their methane emissions (Scope 1) based on direct measurements, as opposed to using generic emission factors. However, reporting at OGMP 2.0 Level 3 involves the development of granular equipment inventory, with the estimation of asset-level emissions through the use of generic, but source-specific emission factors – an important first step – particularly for under-resourced developing countries and which can provide a basis for future measurements to be prioritised. OGMP 2.0 includes not only a company's operated assets, but also its joint ventures (UNEP, 2023[4]) – see Table 2.1 for an overview of OGMP 2.0 Reporting Levels.

OGMP 2.0 has achieved global recognition as a reporting standard that can support robust data collection and reporting for the oil and gas sector across all segments of the value chain. For example, the *Joint Declaration from Energy Importers and Exporters on Reducing Greenhouse Gas Emissions from Fossil Fuels* encouraged the participation of companies in the OGMP 2.0 standard. The EU's regulation on methane emissions reduction in the energy sector also builds on the OGMP 2.0 framework and encourages regulators and third-party verifiers to use the OGMP 2.0 reporting framework, templates and guidance documents as reference criteria. Furthermore, the EU regulation directs the European Commission to develop a methane emissions reporting template taking into account the latest technical guidance documents and reporting templates of the OGMP (European Commission, 2023[5]). Consequently, the global uptake of the OGMP 2.0 framework by jurisdictions and companies can help the EU gain insights ahead of implementation of their own regulatory standards (Olczak, Piebalgs and Balcombe, 2023[6]).

Colombia's 2022 methane regulations set out requirements for companies to monitor methane emissions in accordance with the US EPA's *Method 21 - Volatile Organic Compound Leaks*, as well as an obligation to provide records to the regulator for flared, vented and fugitive emissions. Companies must report to the regulator volumes of flared natural gas on a monthly basis. Companies must quantify volumes of intentionally vented natural gas, either by direct measurement or by mathematical calculation, and report those volumes to the regulator. In addition, operators must identify and quantify the natural gas emissions arising from equipment, components and during well operations.

Box 4.3. USA – Greenhouse Gas Reporting Program

The Greenhouse Gas Reporting Program (GHGRP) requires reporting requirements for a wide range of emission sources and entities across the oil and gas industry. It requires affected owners or operators to collect GHG data, calculate GHG emissions, and follow the specified procedures for quality assurance, missing data, recordkeeping, and reporting.

The GHGRP covers emissions from different aspects of the oil and gas industry through several of its subparts:

- **Subpart W** applies to any **facility that participates in production**, processing, compression, gathering and boosting, storage, transmission, and distribution and that emits 25 000 mt CO_2e or more per year.
- **Subpart Y** applies to any **facility that refines petroleum** and meets the source category definition. This subpart does not specify a minimum level of emissions (or "threshold") reporting requirement, so it applies to all petroleum refineries.
- **Subpart MM** applies to all **petroleum refineries and to importers and exporters of petroleum products and natural gas liquids** whose supplied products would result in 25 000 metric tons CO_2e or more per year if the products were fully combusted.
- **Subpart NN** applies to **suppliers of natural gas liquids and to natural gas distribution companies** that deliver 460 000 Mscf or more of natural gas per year. Facilities subject to Subpart NN requirements must report the emissions that would result from the complete combustion or oxidation of the products that they sell.

The GHGRP sets out different methodologies for measuring emission of GHGs. Facility owners and operators are required to use the methodologies in the relevant subpart (see above) to determine annual emissions for each source category.

Source: (EPA, 2024[7]).

Nigeria's 2022 methane guidelines impose requirements to put in place a system for identifying, classifying, and quantifying methane emissions from process equipment, components and other sources of emissions across oil and gas operations. Section 3.1 provides that the requirement to monitor emissions from operations (flaring and venting) and leaks from equipment (fugitive emissions) applies to all facilities, including well production facilities, natural gas gathering compressor stations, natural gas transmission compressor stations and natural gas processing plants. Operators are directed to undertake measurement/estimation and quantification of GHG emissions (including CO_2, CH_4 and N_2O) from their operations on a quarterly basis.

Governments should require the installation of metering systems for emissions sources where possible (e.g. flow meters on vent heads) while for non-metered sources emissions should be calculated based on emission factors. For example, in Alberta, Directive 060 provides that methane emissions may be quantified using continuous metering, periodic testing, or estimates based on accepted engineering practices. *Directive 060* provides that an operator must include the following information in its annual methane emissions report for each upstream oil and gas facility:

- ***Routinely vented gas*** (gas volumes in m^3 + corresponding mass of methane emitted in kg)
- ***Vented gas from pneumatic instruments and pumps*** (vent gas emitted in m^3 + corresponding mass of methane emitted in kg)
- ***Vented gas from compressor seals*** (vent gas emitted in m^3 + corresponding mass of methane emitted in kg); and
- ***Fugitive emissions*** (in m^3 + corresponding mass of methane emitted in kg).

Additional monitoring and measurement may be required on a case-by-case basis, and in this regard, section 7(1) of the *Alberta Methane Emission Reduction Regulation 244/2018* empowers the regulator to direct an operator to conduct additional monitoring and measurement, related to methane emissions in a manner and frequency specified.

Governments should also be aware of the emergence of MRV frameworks that seek to cover the entire oil and/or gas value chain as these may impact on methane emission reporting in the upstream oil and gas sector. For example, on 15 December 2023, the U.S. Department of Energy announced the creation of an international working group[1] which will develop a consistent framework for the measurement, monitoring, reporting, and verification of methane, carbon dioxide, and other GHG emissions that occur across the *natural gas value chain*. The international working group was formed in recognition that a lack of comparable and reliable information to characterise the GHG intensity of cargos of natural gas limits the ability of buyers to demand and suppliers to provide natural gas with a lower GHG profile, which hinders market-driven emissions abatement efforts (FECM, 2023[8]).

Enhancing accountability through third-party verification and public disclosure

Third party verification refers to an assessment of reported GHG emissions and emissions reductions undertaken by an independent entity. Developing producer countries should ensure that verification requirements are comprehensive and that these requirements go beyond a desktop auditing process of a methodology and include verification of emissions measurements where feasible. Third-party verification is a requirement in Mexico, where oil companies are required to contract the services of an authorised third party to verify the fulfilment of methane emission reductions requirements (IEA, 2021[2]), whereas in Nigeria, data is validated by a multi-disciplinary team comprised of the regulator and the operator. However, the regulator has the discretion to appoint a third-party verifier if required.

Developing producer countries should also take note of third-party verification requirements in the EU Regulation to reduce energy sector methane emissions that apply to imports of oil and gas. From 1 January 2027, importers must demonstrate to EU regulators that oil and gas imported into the EU has been subjected to verification measures at the level of the producer that are equivalent to those that apply to production activities within EU borders. As a result, producing countries seeking to export to the EU will need to ensure that their regulatory framework includes third-party verification provisions that meet the equivalence test (European Commission, 2023[5]). The new EU Regulation to reduce energy sector methane emissions sets out strict requirements for third-party verification for upstream oil and gas projects in the EU. Verifiers must be independent from the operators and are directed to carry out their verification activities in the public interest. In addition, third-party verifiers must be accredited by a national accreditation body (European Commission, 2023[5]).

EFFECT recommends the public disclosure of methane emissions information, including publication on a website. Regulators may choose to publish information on emissions – in a general from or on a company-by-company or project-by-project basis. This approach may create an additional incentive for companies to comply by adding another layer of scrutiny to their operations. In addition, the information generated from MRV requirements can provide governments with increased visibility over the extent of their methane emissions, which in turn can support the design of tailored regulations to further reduce methane emissions for the upstream oil and gas sector.

Several jurisdictions publicly disclose data on methane emissions from flaring and venting on a regular basis – including Brazil, Canada, Guyana, Mexico, and Norway (Ministry of Natural Resources, 2023[9]); (World Bank, 2022[10]). In Alberta and Saskatchewan, the methane emission regulations empower the regulator to publish emissions data on its website and include information on methane emissions at the provincial level but also a breakdown of emissions for individual oil and gas facilities. In Alberta, the Alberta Energy Regulator (AER) publishes an annual report on its website (*ST60B: Upstream Petroleum Industry flaring report*) that includes a list of operators ranked by their flaring and venting emissions. The AER may also make this information available to licensees and operators, in order to encourage and facilitate clustering opportunities, where multiple operators on adjacent fields collaborate to capture associated gas for storage or use. In Saskatchewan, the methane regulations empower the minister to publicly disclose methane emissions data. Section 20 provides that the minister shall publish an annual report on its website setting out the following:

- ***Combined emissions*** at all oil facilities in Saskatchewan;
- ***Combined potential emissions***[2] at all oil facilities in Saskatchewan; and
- ***Individual emissions*** for each oil facility in Saskatchewan.

International transparency initiatives can also play a key role in ensuring public access to methane emissions information. For example, in Nigeria, the Nigeria Extractive Industries Transparency Initiative (NEITI) undertakes a detailed reconciliation process of payments paid for flaring to the Nigerian government. Details of these transactions are published in the NEITI website with a time lag of less than two years (World Bank, 2022[10]).

Notes

[1] International MMRV Working Group participants include Australia, Brazil, Canada, Colombia, East Mediterranean Gas Forum (Observer), Egypt (Observer), European Commission, France, Germany, Italy, Japan, Norway, Republic of Korea, United Kingdom, and the United States.

[2] Potential emissions are calculated in accordance with an emissions intensity limit for each type of crude oil.

References

EPA (2024), *Greenhouse Gas Reporting Program (GHGRP)*, U.S. Environmental Protection Agency, https://www.epa.gov/ghgreporting. [7]

European Commission (2023), *Commission welcomes deal on first-ever EU law to curb methane emissions in the EU and globally*, European Commission, https://ec.europa.eu/commission/presscorner/detail/en/IP_23_5776. [5]

FECM (2023), *DOE Announces Global Collaboration to Reduce Methane Emissions*, U.S. Department of Energy's (DOE) Office of Fossil Energy and Carbon Management (FECM), https://www.energy.gov/fecm/articles/doe-announces-global-collaboration-reduce-methane-emissions. [8]

IEA (2021), *Driving Down Methane Leaks from the Oil and Gas Industry*, International Energy Agency, Paris, https://www.iea.org/reports/driving-down-methane-leaks-from-the-oil-and-gas-industry. [2]

Ministry of Natural Resources (2023), *Data Visualization*, Petroleum Management Programme, https://petroleum.gov.gy/data-visualization?tid=All&tid_1=All. [9]

Mohlin, K. et al. (2022), *Policy instrument options for addressing methane emissions from the oil and gas sector*, Environmental Defense Fund, https://papers.ssrn.com/sol3/papers.cfm?abstract_id=4136535. [3]

OECD (2022), *Equitable Framework and Finance for Extractive-based Countries in Transition (EFFECT)*, OECD Development Policy Tools, OECD Publishing, Paris, https://doi.org/10.1787/7871c0ad-en. [1]

Olczak, M., A. Piebalgs and P. Balcombe (2023), *A global review of methane policies reveals that only 13% of emissions are covered with unclear effectiveness*, One Earth, https://www.sciencedirect.com/science/article/pii/S2590332223001951. [6]

UNEP (2023), *Oil & Gas Methane Partnership 2.0*, United Nations Environment Programme, https://ogmpartnership.com/faq/. [4]

World Bank (2022), *Global Flaring and Venting Regulations: A Comparative Review of Policies*, World Bank, https://thedocs.worldbank.org/en/doc/fd5b55e045a373821f2e67d81e2c53b1-0400072022/global-flaring-and-venting-regulations-a-comparative-review-of-policies. [10]

5 Regulatory building block 2: Setting planning and licencing requirements

Box 5.1. EFFECT recommendations: Planning and licencing requirements

What can governments do?

- Introduce methane reduction requirements in the planning stages of projects, requiring new installations or developments to utilise zero-emitting technologies and have plans in place to capture gas and deliver it to the market
- For new projects, governments should require that field development plans for new oil fields incorporate sustainable utilisation or conservation of the field's associated gas without routine flaring
- Consider introducing methane emission reduction requirements in upstream exploration and production contracts or licences.

Source: (OECD, 2022[1]).

The 2023 Production Gap Report notes that current government plans and projections would lead to an increase in global oil and gas production until at least 2050, resulting in more than double the amount of fossil fuels in 2030 that would be consistent with limiting global warming to 1.5°C (SEI, Climate Analytics, E3G, IISD, UNEP, 2023[2]). Consequently, to keep global climate targets alive, it is imperative for governments planning to develop new oil and gas projects to take urgent steps to reduce methane emissions as early as possible by including methane mitigation requirements in the initial project planning and licencing phase of oil and gas development.

Introducing methane reduction requirements in project planning and licencing

Governments can introduce methane reduction requirements in the planning and licencing stages of projects to ensure that methane emissions are addressed from the outset and ideally before production takes place. The project planning and licensing stage can be an opportunity for governments to require new developments to utilise zero-emitting technologies and have plans in place to capture gas and deliver it to the market. For example, in Kazakhstan, the *Petroleum Law* requires operators to develop a plan to capture and use associated gas before receiving regulatory approval to construct new oilfield projects. These plans must be updated every three years and include one (or more) of the following options:

- associated gas may be used by the operator for onsite power generation
- associated gas may be sold to a third party for processing and marketing
- If processing of the associated gas is uneconomic, the gas may be reinjected into an underground reservoir – either for storage or to maintain reservoir pressure (CLDP, 2023[3]).

Examples of the inclusion of methane reduction requirements in project planning can be found in Malaysia and the United Kingdom. For example, in Malaysia, new oil and gas developments are required to be designed for zero continuous flaring and venting. Field development plans are commonly used in both licencing and contractual regimes and provide an avenue for setting out the conditions under which oil and gas operations should take place – see Box 5.2. In the United Kingdom, the North Sea Transition Authority requires that all field development plans for greenfield oil and gas projects include: zero routine flaring and venting, gas recovery systems, low-carbon electricity options, GHG measurements, and new technologies to reduce emissions (World Bank, 2022[4]).

Box 5.2. Field development plans

Following the delineation of a commercial discovery, a field development plan (FDP) will be required to outline how the oil company intends to develop the petroleum field, manage the impact on the environment and society, as well as forecasts for production and costs. Ideally, the FDP should address every stage of the project's life from initial design, ongoing maintenance and remediation programmes, and finally the decommissioning process.

The FDP is also a useful entry point for regulators to expressly set out methane emission reduction requirements that oil companies must meet during the lifecycle of the field.

Governments should ensure that FDPs include a GHG management plan that specifically covers methane emissions. This could include the requirement to measure emissions from the project, and embed methane mitigation technologies and best practices in the design and operating requirements of the field. A GHG management plan, with specific methane abatement provisions, should include adaptive provisions to allow for the adoption and use of new technologies and practices to further reduce methane emissions within the project area.

FDPs, with their GHG management plans, are a powerful regulatory lever for identifying and mitigating methane emissions in the upstream oil and gas sector emissions. For example, the lack of robust facilities maintenance and remediation is one of the major contributors to petroleum sector emissions in Angola and Nigeria, and an example of where the FDP approval process could be leveraged to lower methane emissions on a site/facility basis.

Source: (AfDB, 2022[5]), (Ogeer, 2022[6]).

References

AfDB (2022), *Minimising Greenhouse Gas Emissions in the Petroleum Sector: The Opportunity for Emerging Producers*, African Development Bank, Abidjan, Côte d'Ivoire, https://www.afdb.org/en/documents/minimising-greenhouse-gas-emissions-petroleum-sector-opportunity-emerging-producers. [5]

CLDP (2023), *Methane Abatement Handbook*, U.S. Department of Commerce, https://cldp.doc.gov/sites/default/files/2023-09/CLDP%20Methane%20Abatement%20Handbook.pdf. [3]

OECD (2022), *Equitable Framework and Finance for Extractive-based Countries in Transition (EFFECT)*, OECD Development Policy Tools, OECD Publishing, Paris, https://doi.org/10.1787/7871c0ad-en. [1]

Ogeer, N. (2022), *Field Development Plans: A Handbook for Government Officials*, Commonwealth Secretariat, https://doi.org/10.14217/comsec.1079. [6]

SEI, Climate Analytics, E3G, IISD, UNEP (2023), *The Production Gap: Phasing down or phasing up? Top fossil fuel producers plan even more extraction despite climate promises*, Stockholm Environment Institute, Climate Analytics, E3G, International Institute for Sustainable Development, United Nations Environment Programme, https://doi.org/10.51414/sei2023.050. [2]

World Bank (2022), *Global Flaring and Venting Regulations: A Comparative Review of Policies*, World Bank, https://thedocs.worldbank.org/en/doc/fd5b55e045a373821f2e67d81e2c53b1-0400072022/global-flaring-and-venting-regulations-a-comparative-review-of-policies. [4]

6 Regulatory building block 3: Setting requirements to reduce flaring and venting

Box 6.1. EFFECT recommendations: Flaring and venting

What can governments do?

- Require that routine flaring at existing oil fields ends as soon as possible, and no later than 2030
- Limit non-routine flaring and venting and clearly define in regulation the circumstances under which operators can flare and vent associated gas without prior approval from the relevant regulatory authority, with reporting requirements and sanctions for non-compliance
- Before granting permission to operators to flare or vent associated gas for economic reasons, require that companies satisfy the regulatory authority that they have investigated all reasonable alternatives to flaring and venting, including reinjection for improved oil recovery or storage, or gas gathering, treatment and sale to downstream energy markets
- Ensure that infrastructure policy is consistent with zero routine flaring and reduced venting objectives and supports the building of pipelines necessary to evacuate gas
- Include dissuasive and proportionate enforcement mechanisms in relevant regulations to deal with non-compliance of flaring and venting of associated gas: for example, penalties and fines, and revocation of the production/operation license
- Ensure that companies implement high flare efficiency ratios to prevent venting of gas
- Identify alternatives to non-routine flaring and venting.

Source: Expanded from (OECD, 2022[1]).

Flaring and venting of natural gas from the upstream oil and gas sector is a significant contributor to global methane and other GHG emissions. Methane emissions primarily arise from venting – i.e. the intentional release of gas into the atmosphere, as well as from flaring – where incomplete combustion of natural gas also releases methane. According to the IEA, the prevention of non-emergency, routine flaring and venting is the most important mitigation measure countries can take as this would avoid almost 20% of oil and gas methane emission globally (IEA, 2023[2]). In fact, the impact of flaring on global methane emissions may be higher than originally thought, as flare efficiency (the percentage of gas that is successfully burned) is often assumed by regulators to be 98%. However, flare efficiency ratios can drop to as low as 10–90% based on impure gas compositions, variable flowrates, remote locations, harsh weather, and poor maintenance practices (Gordon et al., 2022[3]).

Data published by the World Bank's GFMR show that global gas flaring released around 400 million tonnes of CO_2e (including un-combusted methane and black carbon) into the atmosphere in 2021 (International Bank for Reconstruction/The World Bank, 2022[4]). To give an idea of the lost opportunity of capturing these gas resources, the gas that was wastefully flared in 2021, could power all of sub-Saharan Africa (IGSD, 2023[5]).

Policies and regulations to tackle flaring and venting could have a significant effect on the reduction of methane and other GHG emissions globally. For example, the prohibition of venting of natural gas from oil wells could reduce global emissions by 95% (IGSD, 2023[5]). Furthermore, capturing gas that otherwise would be wasted is a cost-effective option: the IEA has estimated that 80% of the options to reduce emissions from oil and gas operations globally could be implemented at no net cost because the cost of the abatement measures are less than the market value of the gas that would be captured – see *Regulatory building block: Incentivising methane emissions abatement*.

Eliminating routine flaring and venting

Governments should put in place policies and regulations to eliminate routine[1] flaring and venting of natural gas. Ideally, such regulations should require that routine flaring at existing oil fields ends as soon as possible, and no later than 2030, in line with international initiatives, including the World Bank's "Zero Routine Flaring by 2030", the Oil & Gas Decarbonization Charter, and the OGCI. For example, New Mexico's Energy, Minerals and Natural Resources Department launched its *Final Natural Gas Waste Reduction Rules* in 2021. These regulations prohibit routine flaring and venting of natural gas in New Mexico and set specific requirements for operators to meet across two phases:

- **Phase 1** – starting October 2021, operators are required to collect and report data that identify potential sources of methane emissions from wellhead to processing sites and beyond. This data will form the basis for establishing individual benchmarks for each operator
- **Phase 2** – operators are required increasing progress toward gas capture targets by meeting a higher level of natural gas capture each year until they reach a 98% capture threshold capture by 2026.

Operators will now be required to pay royalties and taxes on vented and flared volumes, and this includes minor wells as the new regulations apply to all wells in New Mexico, even wells that produce 10 bpd or less. The regulations grant the regulator (the Oil Conservation Commission) with significant powers to enforce compliance, including the ability to deny drilling permits if gas capture targets are not achieved (EMNRD, 2021[6]).

Governments should ensure that regulatory interventions address both flaring and venting as policies that focus only on limiting flaring can result in a corresponding increase in venting – which is more difficult to detect, and which has a greater impact on global warming. For example, such a change in the behaviour of operators has been observed in Turkmenistan following the introduction of a ban on continuous flaring, without tougher parallel policies to limit venting (Olczak, Piebalgs and Balcombe, 2023[7]). In addition, in order to enhance compliance, regulations that impose blanket prohibitions on routine flaring and venting should also provide incentives for that gas to be captured and sold, used on-site, or reinjected for enhanced oil recovery or storage – see *Regulatory building block: Incentivising methane emissions abatement*.

Limiting non-routine flaring and venting

Non-routine flaring and venting refers to exceptional circumstances where flaring and venting may occur in limited volumes – for example, for safety or emergency reasons, well testing or maintenance. In situations where gas cannot be captured, regulations ordinarily favour flaring over venting as combustion (flaring) reduces the methane content of the gas – for example, see Box 6.2 for the approach taken in Nigeria.

Regulations should clearly define the exceptional circumstances under which operators are allowed to flare natural gas. For example, the Canadian province of British Columbia's Drilling and Production Regulation stipulate that a permit holder must not flare gas unless such flaring falls under the below exceptions:

- ***Flaring is required for emergency purposes*** or for drilling operations
- ***Flaring is required for a workover or maintenance*** and the cumulative quantity of gas flared does not exceed 50,000 m^3 in one year (for well sites)
- ***Flaring is required for maintenance purposes*** (at a facility); or
- ***Permission to flare is included in a permit*** (either a well or a facility permit).

Flaring permits can also be used by governments to determine under which circumstances operators can flare and vent associated gas. Flaring permits are required in the Canadian provinces of British Columbia, Saskatchewan (flaring provisions are set out in a facility licence) and Colombia. For example, in Colombia, operators who wish to flare natural gas during initial testing of exploratory and appraisal wells, and for extensive testing, may do so but only in accordance with the terms of a flaring permit. During the production phase, natural gas flaring can only be authorised under the terms of a flaring permit. Flaring permits must be applied for on an annual basis and operators must include an estimation of the volume of gas to be flared, a justification for why such routine flaring is planned, and information on alternatives to flaring (i.e. gas utilisation), where applicable.

Regulations can also set out limits on flaring and venting volumes, prescribe the equipment or process for flaring or venting, and can require operators to apply for flaring permits – which set out specific flaring requirements for each asset in detail.

Box 6.2. Venting requirements in Nigeria

Section 3.3.1 of Nigeria's methane guidelines provides that venting is prohibited in the Nigerian oil and gas industry. However, an operator may be granted waiver to vent natural gas due to operational exigencies. In this case, the following requirements must be met:

- ***Address the root cause of venting*** – the operator shall address the root cause of gas being sent to a cold vent stack, including taking steps to minimise equipment venting from operations and minimising fugitive emissions.
- ***Prioritise flaring where possible*** – any remaining venting shall be routed to a flare, unless the gas mixture is not flammable, or the gas volume/pressure is too small/low for the flare design and thus the flame is not stable.

Source: (NUPRC, 2022[8]).

Non-routine flaring and venting requirements may also differ depending on whether the facility is producing oil or gas. It is more likely for gas production facilities to have infrastructure in place that can more easily accommodate unexpected gas, whereas oil facilities may not be connected to gas pipelines and transportation networks. In situations where gas cannot be captured, regulations ordinarily favour flaring over venting as combustion (flaring) reduces the methane content of the gas. In the Canadian province of Saskatchewan, for example, *Directive PNG036* sets out different requirements for flaring and venting from oil facilities (associated gas) and from natural gas facilities. Oil wells and oil facilities may flare in excess of 900 m^3 per day, whereas flaring at a gas well or gas facility, is not permitted unless it is an emergency and a reasonable level of precaution has been taken to protect human health, public safety, property and the environment and to prevent fire or explosion.

However, there may be exceptional circumstances where regulations allow for the limited venting of associated gas. For example, Colombia's 2022 methane regulations prohibit the venting of associated gas during both the exploration and production phase. In circumstances where associated gas cannot be captured for technical or economic reasons, the gas must be flared. Notwithstanding this prohibition, venting is permitted in the following exceptional circumstances:

- ***Emergency*** – in case of emergency that requires venting of gases into the atmosphere. In this case, the regulator must be informed within 24 hours;
- ***Maintenance*** – due to specific conditions that arise during compliance with the preventive maintenance programme of the facility. In this case, the regulator must be informed within 24 hours.
- ***Pilot light*** – when the volume of vented gas is below that required for a pilot burner to operate.

The British Columbia 2018 flaring guideline sets out requirements for venting that apply to oil facilities processing associated gas. In terms of venting requirements, venting is not an acceptable alternative to flaring, and if gas volumes are sufficient to sustain stable combustion, the gas must be burned or conserved. If venting is the only feasible alternative, it must meet additional requirements – see Box 6.3.

Box 6.3. British Columbia: Regulatory requirements for venting

Chapter 7: Venting and Fugitive Emissions Management Requirements

Venting is not an acceptable alternative to conservation or flaring. Venting is the least preferred option and gas should be flared under all except the most exceptional circumstances.

Venting requirements:

- All continuous and temporary venting and their sources must be evaluated using a vent evaluation decision tree
- Permit holders must burn all non-conserved volumes of gas if volumes and flow rates are sufficient to support stable combustion
- Vented gas must not constitute a safety hazard
- The quantity and duration of vented gas must be minimised
- A permit holder must have an adequate programme for managing fugitive emissions.

Source: (BC Oil & Gas Commission, 2018[9]).

Ensuring high flare efficiency ratios

Flaring refers to a process where natural gas is combusted and the methane component of the gas is destroyed, resulting in emissions of CO_2 into the atmosphere. Flare efficiency ratios represent an underappreciated methane source and mitigation opportunity. Industry and governments generally assume that the flare efficiency (the percentage of gas that is successfully burned) is around 98% and therefore that the methane released from flaring remains a marginal amount. However, recent studies have indicated that global flare efficiency ratios may be much lower than originally assumed and this may have a substantial impact on global methane emissions. Estimates[2] from the IEA and academic research indicate a global average combustion efficiency of around 92% and that the incomplete combustion of gas from flares causes around 10% of total methane emissions from oil and gas operations (IEA, 2024[10]); (Plant et al., 2022[11]).

Although flares should always be lit and well maintained, this is not always the case as flares can become temporarily unlit, for example, due to strong winds, a pilot flame malfunction, or low-quality gas "snuffing" the flare (World Bank, 2024[12]). The scale of the issue of unlit flares may also be underappreciated. For example, a recent study by the Environmental Defense Fund in the Permian Basin found that 11% of flares were either unlit or malfunctioning and therefore were venting methane to the atmosphere. Furthermore, unlit flaring may continue for a longer period of time at unmanned facilities if not detected or until sufficient resources are deployed to reignite the flare (EDF, 2021[13]).

Consequently, governments should ensure that regulations set sufficiently high flare efficiency ratios and provide for the on-going monitoring of flaring to ensure compliance. For example, Nigeria's methane guidelines specify that all flared gas shall be combusted with an auto-igniter or continuous pilot light and a design destruction removal efficiency of at least 98% for hydrocarbons (NUPRC, 2022[8]). The EU's methane regulation stipulates that where new flare stacks are built, combustion devices with an auto-igniter or continuous pilot must have a destruction and removal design efficiency of at least 99% (European Commission, 2023[14]). Governments may also consider the use of automated monitoring systems of flare stacks, in order to avoid any operational disruptions caused by direct physical inspections.

Identifying alternatives to non-routine flaring and venting

Improvements to combustion efficiency or capturing gas that would otherwise be flared can deliver significant decarbonisation results. If the captured gas is used to displace the usage of more carbon-intensive energy (for example, coal or heavy fuel oil), the greenhouse gas impact can be even greater (CATF, 2023[15]). Consequently, regulations should specify that companies must satisfy the regulatory authority that they have investigated all reasonable alternatives to non-routine flaring and venting, before being granted permission to flare or vent associated gas for economic reasons. For example, in British Columbia, the flaring guideline sets out clear requirements for gas conservation at oil facilities processing associated gas – see Box 6.4.

Box 6.4. British Columbia: Requirements to conserve natural gas

Oil facilities processing associated gas

Gas conservation is expected at all new oil facilities. However, new oil facilities where conservation is not economic or practical may be approved by the regulator on a site-by-site basis. If the net present value (NPV) of the gas conservation project is greater than CAD 50 000, the wells should be shut in until conservation is implemented. For existing oil sites, operators should conserve associated gas where:

- ***Combined flaring and venting volumes*** are greater than 900 m^3/day per site and the decision tree process and economic evaluation result in a NPV of greater than CAD 50 000
- ***The gas to oil ratio (GOR)*** is greater than 3 000 m^3/m^3. All wells producing with a GOR greater than 3 000 m^3/m^3 at any time during the life of the well should be shut-in until the gas is conserved
- ***Flared volumes*** are greater than 900 m^3/day per site and the flare is within 500 m of an existing residence, regardless of economics.

Flaring and venting at natural gas facilities

Operators must conserve all gas that is economic to conserve, although there are also exceptions for safety and environmental reasons. Section 3.2 provides that operators must conserve gas at natural gas facilities where:

- ***Conservation economics*** produce a NPV greater than CAD 0
- ***Flared volumes*** are greater than 4 000 m³/day per site and the flare is within 1 000 m of an existing residence.

Source: (BC Oil & Gas Commission, 2018[9]).

Conservative measures that provide alternatives to flaring and venting may include reinjection for improved oil recovery or storage, or gas gathering, treatment and sale to downstream energy markets. However, the reinjection of associated gas for improved oil recovery may lead to increased GHG emissions due to the additional production and subsequent consumption of that oil. Regulations should also encourage greater co-ordination and collaboration between industry participants, especially where capturing gas from flares of one operator requires access to critical processing facilities and/or pipelines of another operator (CATF, 2023[15]). Indeed, the location and availability of infrastructure can be crucial to the commerciality of flare capture projects as demonstrated by the recent reduction in gas flaring in Egypt.

Egypt's flaring intensity is two times higher than the world average. However, in recent years, Egypt has significantly reduced flaring, by an average of 6% per year over the last six and flaring volumes are now 26% below 2016 levels. Flare capture projects have contributed to lowering Egypt's flaring intensity and are driven by the fact that 75% of flared volumes take place within 20 km of an existing gas pipeline. In other cases, operators have made use of captured gas for on-site power generation. For example:

- ***Construction of a new pipeline*** – following the discovery at the Ash oil field, the operator, United Oil and Gas, installed a 20 km pipeline to link the field with the (existing) El Salmiya gas processing facility. The field produces 5 million scf/day of associated gas, demonstrating that a moderate-length pipeline can be a commercially viable option, even for relatively small flares.
- ***Gas recovery for power generation*** – Operators Pharos Energy and Apache developed flare-to-power projects to capture and use 1 million scf/day and 3 million scf/day of associated gas respectively from their operations. In both cases, the power generated by the captured gas displaces diesel – which not only is more polluting but is in short supply in Egypt. The CATF has estimated that these projects saved up to 3 million litres of diesel per month – reducing emission and also lowering operating costs by several tens of million dollars per year (CATF, 2023[15]).

Another alternative to non-routine flaring and venting is where the government (rather than the operator) has the power to access and utilise associated gas. For example, in Nigeria, the federal government has the right to take any associated gas that would have been flared either free of cost at the flare or at a cost agreed with the operator. In addition, through the Nigerian Gas Flare Commercialisation Programme (NGFCP), the government can allocate rights to third parties to monetise associated gas at specific flaring sites through a competitive bidding process. The NGFCP was developed to tackle small flaring sites that proved more difficult to monetise – for further analysis of the NGFCP see *Regulatory building block: Incentivising methane emissions abatement*.

Notes

[1] Routine flaring and venting occur when oil field operators opt to burn the "associated" gas that accompanies oil production, or simply release it to the atmosphere, rather than to build the equipment and pipelines to capture it (IEA, 2023[16]).

[2] These estimates include both unlit flares and inefficient combustion, and are based on bottom-up assessment of production types; facility and flare design practices; operators; changes in produced volumes over field lifetime; local crosswind variability; and the strength of regulation, oversight and enforcement in flaring sites around the world.

References

BC Oil & Gas Commission (2018), *Flaring and Venting Reduction Guideline*, BC Oil & Gas Commission, https://www.bc-er.ca/node/5916/. [9]

CATF (2023), *Leadership on flaring in Egypt: Recent successes and future opportunities in the lead-up to COP27*, Clean Air Task Force (CATF), https://www.catf.us/resource/leadership-flaring-egypt-recent-successes-future-opportunities-lead-up-cop27/. [15]

EDF (2021), *PermianMAP Final Report*, Environmental Defence Fund (EDF), https://blogs.edf.org/energyexchange/wp-content/blogs.dir/38/files/2022/11/PermianMAPFinalReport.pdf. [13]

EMNRD (2021), *Final Natural Gas Waste Reduction Rules*, Energy, Minerals and Natural Resources Department, https://www.emnrd.nm.gov/wp-content/uploads/sites/6/OCDMethaneRuleReleaseMarch252021.pdf. [6]

European Commission (2023), *Commission welcomes deal on first-ever EU law to curb methane emissions in the EU and globally*, European Commission, https://ec.europa.eu/commission/presscorner/detail/en/IP_23_5776. [14]

Gordon, D. et al. (2022), *Know Your Oil and Gas: Generating Climate Intelligence to Cut Petroleum Industry Emissions*, Rocky Mountain Institute, https://rmi.org/insight/kyog/. [3]

IEA (2024), *Global Methane Tracker 2024*, International Energy Agency, Paris, https://www.iea.org/reports/global-methane-tracker-2024. [10]

IEA (2023), *Gas Flaring (website)*, https://www.iea.org/energy-system/fossil-fuels/gas-flaring. [16]

IEA (2023), *Global Methane Tracker 2023*, International Energy Agency, Paris, https://www.iea.org/reports/global-methane-tracker-2023. [2]

IGSD (2023), *A Primer on Cutting Methane: The Best Strategy for Slowing Warming in the Decade to 2030*, Institute for Governance & Sustainable Development, https://www.igsd.org/wp-content/uploads/2022/09/IGSD-Methane-Primer_2022.pdf. [5]

International Bank for Reconstruction/The World Bank (2022), *Global Flaring and Venting Regulations: A Comparative Review of Policies*, World Bank Publications, https://thedocs.worldbank.org/en/doc/fd5b55e045a373821f2e67d81e2c53b1-0400072022/original/Global-Flaring-and-Venting-Regulations-A-Comparative-Review-of-Policies.pdf. [4]

NUPRC (2022), *Guidelines for Management of Fugitive Methane and Greenhouse Gases Emissions in the Upstream Oil and Gas Operations in Nigeria*, Nigerian Upstream Petroleum Regulatory Commission, https://www.nuprc.gov.ng/wp-content/uploads/2022/11/METHANE-GUIDELINES-FINAL-NOVEMBER-10-2022.pdf. [8]

OECD (2022), *Equitable Framework and Finance for Extractive-based Countries in Transition (EFFECT)*, OECD Development Policy Tools, OECD Publishing, Paris, https://doi.org/10.1787/7871c0ad-en. [1]

Olczak, M., A. Piebalgs and P. Balcombe (2023), *A global review of methane policies reveals that only 13% of emissions are covered with unclear effectiveness*, One Earth, https://www.sciencedirect.com/science/article/pii/S2590332223001951. [7]

Plant, G. et al. (2022), *Inefficient and unlit natural gas flares both emit large quantities of methane*, Science, https://www.science.org/doi/10.1126/science.abq0385. [11]

World Bank (2024), *Global Gas Flaring Tracker Report 2024*, International Bank for Reconstruction and Development / The World Bank, https://www.worldbank.org/en/programs/gasflaringreduction/publication/2024-global-gas-flaring-tracker-report. [12]

7 Regulatory building block 4: Addressing fugitive methane emissions

Box 7.1. EFFECT recommendations: Fugitive methane emissions

What can governments do?

- Require over time the replacement of equipment that is designed to vent methane with technologies that are zero emitting or the use of vapour collection to reroute vented methane back into the pipeline.
- Design robust leak detection and repair programmes (LDAR) based on environmental outcomes that emphasise repairing detected leaks and preventing leaks.
- Require detection campaigns with specified frequency (e.g. quarterly), and specify equipment to be used, detection thresholds and time limits for repairs. Using methane regulations allows for the use of alternative leak detection and repair programmes that can achieve equivalent outcomes.

Source: (OECD, 2022[1]).

Fugitive emissions (also referred to as "leaks") are the most significant source of methane emissions from the oil and gas sector (Mohlin et al., 2022[2]), and are widely recognised as both a source of global GHG emissions and of local air pollution. Fugitive emissions are caused by malfunctioning or poorly maintained natural gas infrastructure – primarily, improperly fitted connection points or deteriorating seals; and the changes in pressure, temperature, or mechanical stresses that can lead to equipment or component degradation (European Commission, 2019[3]). Consequently, regulatory measures to address fugitive methane emissions are critical to lowering emissions and methane intensity, throughout the value chain and, in particular in the upstream oil and gas sector.

Equipment and technology standards

In line with EFFECT recommendations, governments can set out equipment and technology standards that operators must meet during the oil and gas production process. Such requirements can be prescriptive or performance-based or a combination of both:

- **Prescriptive instruments** direct regulated entities to undertake or not to undertake specific actions or procedures (IEA, 2021[4]). For example, these can require operators to replace a device with an upgraded version that emits less methane or install new equipment that recovers emissions.
- **Performance-based instruments** establish a mandatory performance standard for regulated entities but do not dictate how the target must be achieved (IEA, 2021[4]). For example, defining an emissions intensity benchmark (amount of emissions per unit of oil or gas production), or specifying a percentage of gas (from total gas production) that must be captured and/or used (Mohlin et al., 2022[2]).

The use of either prescriptive or performance-based regulatory instruments will vary across jurisdictions depending on the nature of the industry and capabilities and resources of regulatory agencies. On the one hand, **prescriptive instruments** are not based on emission quantification and therefore, the required technologies can be put in place before methane quantification methods have been implemented (Mohlin et al., 2022[2]). As such, they are less difficult to enforce as compliance is linked to equipment used at a facility. However, the prescriptive approach is by its nature less flexible and less able to respond to rapidly evolving technologies. For example, Nigeria's 2022 methane guidelines impose a number of equipment-specific standards on operators in the upstream oil and gas sector applicable to new and existing oil and gas facilities. Section 3.4.1 sets out requirements for centrifugal compressors, and specifies that operators route oil degassing unit emissions either to a vapor recovery or a combustion device. Alternatively, operators can design/retrofit the compressor using dry seals. In addition, in order for prescriptive regulations to be effective, governments will need to undertake sufficient upfront investigation to understand where emissions are coming from in order for the regulations to be correctly targeted.

On the other hand, **performance-based instruments** are easier to develop and allow more flexibility for operators to choose methods of compliance. They are agnostic as to the use of specific technology. For this reason, they support increased technology innovation and can prove more cost-effective as they incentivise operators to deploy latest available equipment and technologies and achieve faster methane emissions reductions. However, performance-based instruments rely on accurate emission monitoring, which may limit their application in jurisdictions with lower capacities due to the expertise needs and time required to ensure proper enforcement and compliance (Olczak, Piebalgs and Balcombe, 2023[5]).

Equipment and technology standards target high-emitting oil and gas equipment by requiring the installation or replacement of components and devices to meet methane emission standards. The adoption of best available equipment and technology can lead to significant reductions in methane emissions (as well as other GHGs and volatile organic compounds (VOCs)) due to their performance relative to existing equipment and components. For example, section 5(1) of the 2018 Canadian federal methane regulations provides that gas conservation equipment that is used at an upstream oil and gas facility must be operated in such a manner that at least 95% of the hydrocarbon gas that is routed to the equipment is captured and conserved.

Phasing-in zero emitting technologies

Governments should ensure that regulations require the replacement of old equipment with technologies that are zero emitting. For example, the EU's *Regulation to reduce energy sector methane emissions* provides that where a site is built, replaced or refurbished in whole, operators shall utilise only commercially available zero-emitting pneumatic devices, compressors, atmospheric pressure storage tanks, sampling and measurement devices and dry gas seals (European Commission, 2023[6]). In another example, the U.S. EPA's *Final Rule to Reduce Methane and Other Harmful Pollution from Oil and Natural Gas Operations* requires pneumatic controllers to meet a methane and VOC emission rate of zero. In Mexico, article 41 of the 2018 Mexico methane regulations provides that, during the design phase of new oil and gas facilities, the operator must select pneumatic pumps driven by compressed air or electric pumps. In British Columbia, regulatory provisions encourage the replacement of pneumatic pumps over time – see Box 7.2.

Box 7.2. British Columbia: Regulatory requirements for pneumatic pumps

Section 52.06 of the 2010 British Columbia Drilling and Production Regulation

(1) A facility permit holder who operates a facility must not use at the facility a pneumatic pump that emits natural gas unless the pump:

- (a) was installed before January 1, 2021; or
- (b) is operated 750 hours or less per year.

(2) Beginning on January 1, 2021, a facility permit holder who operates a facility that uses a pneumatic pump that emits natural gas must maintain a record that contains:

- (a) description of the pump and the type of fluid pumped, and
- (b) the following information for each calendar month:
  - (i) the number of hours the pump is operated
  - (ii) the volume of natural gas emitted from the pump.

Source: (Government of British Columbia, 2022[7]).

Governments adopting new methane regulations may opt for equipment and technology standards to be phased in over time to allow operators time and flexibility to adjust to these new requirements. For example, the U.S. EPA's *Final Rule* provides a one-year phase-in for zero-emissions standards for new process controllers and most new pumps outside of Alaska in order to provide industry time to prepare to meet requirements and to secure necessary equipment (EPA, 2023[8]). Facilities in Alaska where access to electrical power from the power grid is not available are exempted from these requirements (EPA, 2023[9]). Regulations may also specify different equipment and technology standards for new and existing facilities. For example, Colombia's methane regulations require that for new facilities, where reciprocating compressors are used, they must be connected to gas and vapor conservation equipment when technical conditions allow it, or to a controlled burning system in order to capture amounts of gas that are vented from reciprocating compressors during normal operations. Whereas, for existing facilities with centrifugal compressors with wet seals, operators must redirect emissions during degassing of wet seals to a vapor conservation equipment or replace wet seals with dry seals. Similarly, for pneumatic controllers, during the design phase of new facilities, operator should select pneumatic controllers that use compressed air rather than natural gas and/or are powered by electricity. Whereas, for existing facilities with pneumatic controllers powered by natural gas, operators should substitute the use of natural gas for compressed air or replace pneumatic controllers with mechanical or electricity ones.

In Nigeria, the methane guidelines prohibit the use of natural gas-driven pneumatic controllers at grid-connected facilities due to their leak potential and stipulates that operators shall instead retrofit facilities with zero bleed controllers, including controllers powered by electricity or instrument air or where emissions are directed to a vapor recovery system for capture. However, if these options are not feasible, the regulations provide that operators may use a flare, and production facilities that do not have access to grid-electricity are also exempted from these requirements. Nigeria's methane guidelines provide a 5-year phase-in period for existing well production facilities without grid access:

- **Within one year**, an operator shall ensure that 25% of pneumatic controllers are zero-bleed controllers, and the remainder are low bleed[1]
- **Within two years**, the operator shall ensure that 65% of pneumatic controllers are zero-bleed controllers, and the remainder are low bleed
- **Within three years**, the operator shall ensure that 75% of pneumatic controllers are zero-bleed controllers, and the remainder are low bleed
- **Within four years**, the operator shall ensure that 85% of pneumatic controllers are zero-bleed controllers, and the remainder are low bleed
- **Within five years**, the operator shall ensure that all pneumatic controllers are zero bleed controllers (NUPRC, 2022[10]).

Leak detection and repair regulatory requirements

Alongside equipment and technology standards, governments should set out robust and effective leak detection and repair (LDAR) programmes to identify and address sources of fugitive methane emissions as they arise. The main elements of a robust and effective LDAR programme include:

- **Scope of facilities to be inspected** – LDAR requirements may specify which facilities and equipment are subject to inspection requirements
- **Authorised LDAR equipment and technologies** – LDAR requirements may direct operators to use a combination of traditional on-site techniques (e.g. optical gas imaging) or advanced techniques – including aircraft, unmanned aerial vehicles (drones), mobile ground labs, and satellites
- **Frequency of inspections** – the more frequent that monitoring is carried out (by the operator), the more effective the LDAR programme will be in identifying and repairing sources of fugitive methane emissions (Mohlin et al., 2022[2])
- **Detection threshold and repair requirements** – this may include a leak threshold that triggers the obligation to repair and the allowance timeframe for carrying out those repairs
- **Reporting and recordkeeping** – LDAR requirements may require companies to keep records of their surveys, detected leaks, and repair actions, and to report to competent authorities (CLDP, 2023[11]).

Governments should design robust LDAR programmes based on environmental outcomes that emphasise repairing detected leaks and preventing leaks. For example, in British Columbia, the 2018 Flaring and Venting Reduction Guideline provides that a facility permit holder must have an adequate fugitive emissions management programme. In addition, permit holders must develop and implement a programme to detect and repair leaks, and that these programmes should meet or exceed the *Canadian Association of Petroleum Producer's Best Management Practice for Fugitive Emissions Management*. In another example, the 2018 Canadian federal methane regulations set out requirements directing operators to carry out LDAR programmes in order to limit fugitive emissions containing hydrocarbon gas from equipment components at oil and gas facilities.[2]

There are a number of different technologies and equipment that operators can use to detect methane emissions, and new technologies are developing rapidly. The International Association of Oil & Gas Producers (IOGP), an upstream oil and gas industry body, has created a filtering tool to assist operators to identify appropriate methane detection and quantification technology – see Box 7.3.

Box 7.3. Methane detection and quantification technology filtering tool

Several technologies are available to help producers detect and quantify methane emissions, a prerequisite to leak repair. To assist in narrowing down technologies appropriate for a particular site or operation, the IOGP, in partnership with other industry associations – OGCI and Ipieca, has developed a publicly available filtering tool.

This online technology filtering tool features detailed technology data sheets covering over 50 technologies, and decision trees to guide technology deployment.

Using a checkbox method, producers input factors such as the accessibility of the site, the characteristics of the area wherein the site is located, and the required sensitivity of the equipment. The database is non-exhaustive, but it is regularly updated.

Source: (IOGP, 2024[12]).

LDAR campaigns – frequency, thresholds and repairs

Regulatory frameworks should require operators to undertake detection campaigns with specified frequency, detection thresholds and time limits for repairs. More frequent surveys lead to faster detection and repair of methane leaks, but this comes with an added cost of compliance. According to the IEA, the most typical frequency for LDAR campaigns is quarterly (IEA, 2021[4]). Table 7.1 provides an overview of LDAR frequency, detection thresholds and repairs timeframes in selected jurisdictions.

Table 7.1. Regulatory requirements for leak detection and repair programmes

LDAR frequency, detection thresholds and repairs timeframes in selected jurisdictions

	Annual frequency of LDAR inspections	Leak threshold for repair	Timeframe for repair following detection
Canada (Federal)	3	500 ppm	30 days*
Canada (Alberta)	1 or 3	10 000 ppm	24 hours (safety issues); or 30 days*
Canada (British Columbia)	1 or 3	500 ppm	30 days**
Canada (Saskatchewan)	2	500 ppm	30 days*
European Union	***	500 – 7 000 ppm	30 days
Colombia	2	500 ppm	30 days**; or12 months (offshore)
Mexico	4	500 ppm	1, 3 or 15 days* (depending on leak size)
Nigeria	4	500 ppm	5 working or 14 days* (depending on leak size)
USA (Federal)	4-6	500 ppm	30-60 days
USA (Colorado)	1-12****	500 ppm	5 days (attempt to repair), 30 days (complete repair)

Note: * If the repair can be carried out while the equipment component is operating. If not, then repair must be made during the next planned shutdown.
** If the repair can be carried out while the equipment component is operating. If not, then repair must be made when the volume of natural gas from all leaks is greater than the volume of natural gas that would be released if the operating equipment is removed.
*** LDAR inspections differ according to the component and type of material – from 3-36 months.
**** LDAR inspections differ depending on magnitude of emissions.
Source: Authors' compilation.

In the European Union, the inspection frequency differs according to the location of the specific component (i.e. onshore, offshore, underground) and type of material (i.e. asbestos, copper, protected steel). Similarly, the leak threshold for repair[3] is determined by reference to the location of the component.

- 500 parts per million in volume of methane or 1 gram per hour of methane for aboveground components and for offshore components above the sea level
- 1 000 parts per million in volume of methane or 5 grams per hour of methane for the second step of underground components
- 7 000 parts per million in volume of methane or 17 grams per hour for offshore components below the sea level and below the seabed (European Commission, 2023[6]).

Box 7.4. Alternative LDAR survey frequencies in the European Union

Article 14 of the EU's *Regulation to reduce energy sector methane emissions* provides that where operators producing or processing natural gas or oil can demonstrate that during the five preceding years less than 1% of all their components in each site are leaking and that the methane emissions associated with these leaks aggregated represent less than 0.08% of the total volume of gas or 0.015% of the total mass of oil processed or extracted, then different LDAR survey frequencies may be used.

Type 1 and 2 LDAR surveys

Operators shall repair or replace all components found to be emitting methane at or above the following levels at standard temperature and pressure and using detection devices in accordance with the manufacturer specifications for operation and maintenance:

(a) for type 1 LDAR surveys: 7 000 parts per million in volume of methane or 17 grams per hour of methane

(b) for type 2 LDAR surveys:

i. 500 parts per million in volume of methane or 1 gram per hour of methane for aboveground components and for offshore components above the sea level

ii. 1 000 parts per million in volume of methane or 5 grams per hour of methane for the second step of LDAR surveys of underground components

iii. 7 000 parts per million.

Alternative LDAR survey frequencies

These alternative LDAR survey frequencies are subject to the approval of the relevant authorities and must comply with the following requirements:

- For all components at processing locations, Type 1 LDAR surveys are performed at least every 12 months
- For at least 25% of all components at processing locations, Type 2 LDAR surveys are performed every 12 months, ensuring that all components are checked every 48 months
- For all components at production locations, Type 1 LDAR surveys are performed at least every 36 months
- For all components at production locations, Type 2 LDAR surveys are performed at least every 60 months.

Source: (European Commission, 2023[6]).

A similar approach is taken in the United States where inspection frequency and repairs timeframes differ according to the nature of the facility and specific LDAR activity. For example, quarterly audio, visual, and olfactory (AVO) monitoring surveys are required at small well sites and sites with a single wellhead. If a leak is detected, an initial attempt at repair must be made within 15 days after detecting fugitive emissions, and the repairs must be completed within 15 days after first attempt. For multi-wellhead sites (without major production and processing equipment), quarterly AVO monitoring surveys are required and if leaks are detected, initial attempt at repair must be made within 15 days, and repairs must be completed within a further 15 days. In addition, semi-annual optical gas imaging (OGI) surveys are required. Should a leak be detected, an initial attempt at repair must be made within 30 days after detecting fugitive emissions, and the repairs must be completed within 30 days after first attempt (EPA, 2023[9]).

In Canada, the 2018 Canadian federal regulations require inspections at least three times per year and at least 60 days after a previous inspection. This requirement (to limit inspection to three times per year), reflects the environment of the Canadian upstream oil and gas sector where LDAR inspections are not possible in offshore and arctic locations during winter months (IEA, 2021[4]). A release of hydrocarbons from an equipment component is considered a leak if the release consists of at least 500 parts per million by volume (ppmv) of hydrocarbons, as determined by an inspection conducted by means of an eligible portable monitoring instrument in accordance with EPA Method 21.[4] If a LDAR inspection identifies an actionable leak, a repair must be made within 30 days if the repair can be carried out while the equipment component is operating, or if not, during the next planned shutdown.

In Colombia, article 42 of the 2022 Colombia methane regulations provides that the operator must carry out LDAR activities at its oil and gas facilities. Although there is an exemption for facilities that operate with a potential for emissions or leaks of less than 60 000 standard m^3 per year. Inspections are required twice per year, and a leak is successfully repaired when it is reduced to less than 500 ppm or when the instruments used for detection do not detect visible emissions. In Mexico, article 71 of the 2018 Mexico methane regulations directs operators to carry out a LDAR programme for each project on a quarterly basis. The threshold for a leak that triggers a repair obligation in 500 ppm.

In Nigeria, the frequency for LDAR inspections is phased in over three years. In the first year after implementation of the Nigeria methane guidelines, an operator shall conduct one inspection at each facility, in the second year, two inspections are required, and in the third and subsequent years, three inspections are required. A leak is successfully repaired when it is reduced to less than 500 ppm (using EPA Method 21) or when an infra-red camera or any other detection technology approved by the regulator does not detect emissions. Repair obligations differ according to the size of the leak and/or if the component cannot be repaired without a shutdown:

- ***Larger leaks*** (50 000 ppmv) shall be repaired within 5 working days of discovery
- ***Small leaks*** (5 000 ppmv) shall be repaired within 14 days of discovery.

The Nigeria methane guidelines note that if the relevant component is a critical one that cannot be repaired without shutdown, operators shall minimise the leak within one day of detection and repair the leak by the end of the next planned process shutdown or within one year, whichever is sooner.

Regulations may provide exemptions or delays to repair timeframes due to the availability or access to certain equipment or components. For example, in response to supply chain concerns, the U.S. EPA's *Final Rule to Reduce Methane and Other Harmful Pollution from Oil and Natural Gas Operations* allows operators additional time to repair fugitive emission components if a replacement is required but parts cannot be acquired or installed due to the following conditions:

- Replacement valve supplies have been sufficiently stocked but are depleted at the time of the repair
- Replacement fugitive emissions component (or a part) requires custom fabrication (EPA, 2023[9]).

LDAR campaigns – equipment specifications

Regulations should specify the equipment to be used in LDAR campaigns. For example, the U.S. EPA's *Final Rule* specifies the use of detection instruments to be used at different facilities – small well sites and sites with a single wellhead (AVO), multi-wellhead sites, sites with major production and processing equipment, and compressor stations (AVO and OGI). In Nigeria, section 3.2.1 of the 2022 Nigeria methane guidelines specifies that LDAR must be conducted using OGI, laser beam technology or any other mature technology approved by the regulator. In Colombia, detection activities must be carried out using: OGI instruments, laser leak detectors, soap solution detection, organic vapor analysers or toxic vapor analysers, acoustic leak detection, electronic gas detectors, or any other that is authorised by the regulator.

As recommended in EFFECT, governments may also consider providing regulatory flexibility for operators to deploy alternative technologies for leak detection and repair programmes that can achieve equivalent outcomes to reflect technological developments. This approach has been implemented in the United States and the Canadian province of Alberta. For example, in recognition of the rapid and continued advancement of technologies, the U.S. EPA's *Final Rule* allows operators, technology developers and other entities to seek approval for the use of new methane detection technologies to supplement their existing ground based OGI surveys and AVO inspections, and to ensure that the regulations keep up with the pace of innovation in the sector. The EPA will assess technology approval requests within 90 days and will issue an approval or disapproval within 270 days (EPA, 2023[8]).

In the Canadian province of Alberta, *Directive 060* provides that the Alberta Energy Regulator will consider innovative and science-based alternatives to the standard fugitive emissions management programme. Such alternative programmes may incorporate technologies such as unmanned aerial vehicles, vehicle-mounted sensors, and continuous monitoring devices to detect, track, repair, and report fugitive emissions. Operators who wish to use an alternative fugitive emissions management programme must first submit a proposal to the Alberta Energy Regulator for approval.

Notes

[1] Emitting less than 0.17 standard cubic meters per hour of natural gas.

[2] Applicable to upstream facilities that produce or receive at least 60 000 standard m^3 of hydrocarbon gas per year.

[3] Applicable to type 2 leak detection and repair surveys.

[4] Two methodologies are ordinarily used to detect leaking equipment in LDAR programmes: 1) *EPA Method 21* developed by the U.S. Environmental Protection Agency uses a hydrocarbon ionisation detector, and is a widely accepted method; and 2) *Optical Gas Imaging (OGI)* uses an infra-red camera – this is a more recent methane detection technique but is gaining increasing acceptance.

References

CLDP (2023), *Methane Abatement Handbook*, U.S. Department of Commerce, https://cldp.doc.gov/sites/default/files/2023-09/CLDP%20Methane%20Abatement%20Handbook.pdf. [11]

EPA (2023), *EPA's Final Rule for Oil and Natural Gas Operations*, U.S Environmental Protection Agency, https://www.epa.gov/controlling-air-pollution-oil-and-natural-gas-operations/epas-final-rule-oil-and-natural-gas. [8]

EPA (2023), *EPA's Final Rule for Oil and Natural Gas Operations Will Sharply Reduce Methane and Other Harmful Pollution.*, United States Environmental Protection Agency, https://www.epa.gov/controlling-air-pollution-oil-and-natural-gas-operations/epas-final-rule-oil-and-natural-gas. [9]

European Commission (2023), *Commission welcomes deal on first-ever EU law to curb methane emissions in the EU and globally*, European Commission, https://ec.europa.eu/commission/presscorner/detail/en/IP_23_5776. [6]

European Commission (2019), *Best available techniques guidance document on upstream hydrocarbon exploration and production – Final guidance document*, European Commission, https://data.europa.eu/doi/10.2779/607031. [3]

Government of British Columbia (2022), *BC Drilling and Production Regulation (B.C. Reg. 282/2010)*, Government of British Columbia, https://www.bclaws.gov.bc.ca/civix/document/id/crbc/crbc/282_2010. [7]

IEA (2021), *Driving Down Methane Leaks from the Oil and Gas Industry*, IEA, Paris, https://www.iea.org/reports/driving-down-methane-leaks-from-the-oil-and-gas-industry. [4]

IOGP (2024), *Methane detection and quantification technology filtering tool*, International Association of Oil & Gas Producers (IOGP), https://www.iogp.org/workstreams/environment/environment/methane-emissions-detection-and-quantification/methane-detection-and-quantification-technology-filtering-tool/tool/. [12]

Mohlin, K. et al. (2022), *Policy instrument options for addressing methane emissions from the oil and gas sector*, Environmental Defense Fund, https://papers.ssrn.com/sol3/papers.cfm?abstract_id=4136535. [2]

NUPRC (2022), *Guidelines for Management of Fugitive Methane and Greenhouse Gases Emissions in the Upstream Oil and Gas Operations in Nigeria*, Nigerian Upstream Petroleum Regulatory Commission, https://www.nuprc.gov.ng/wp-content/uploads/2022/11/METHANE-GUIDELINES-FINAL-NOVEMBER-10-2022.pdf. [10]

OECD (2022), *Equitable Framework and Finance for Extractive-based Countries in Transition (EFFECT)*, OECD Development Policy Tools, OECD Publishing, Paris, https://doi.org/10.1787/7871c0ad-en. [1]

Olczak, M., A. Piebalgs and P. Balcombe (2023), *A global review of methane policies reveals that only 13% of emissions are covered with unclear effectiveness*, One Earth, https://www.sciencedirect.com/science/article/pii/S2590332223001951. [5]

8 Regulatory building block 5: Incentivising methane emissions abatement

Box 8.1 EFFECT recommendations: Economic instruments and incentives to reduce methane emissions

What can governments do?

- Implement a reward-penalty system to incentivise methane emissions reduction
- Ensure that the ownership of associated gas is clear and understand associated implications
- Address gas infrastructure challenges
- Ensure that methane investments are cost-recoverable
- Encourage associated gas clustering opportunities
- Prioritise access for associated gas into the gas network.

Source: Adapted from (OECD, 2022[1]).

Compliance with measurement, monitoring, reporting, verification and other methane abatement requirements can imply higher operating and capital costs, making oil and gas projects less profitable. Such additional costs are the price of negative externalities of oil and gas projects on the environment that need to be factored into project feasibility studies. In practice, incremental costs can deter both governments and operators from delivering deep emissions cuts.

Governments should consider providing incentives for companies to reduce methane emissions from upstream oil and gas operations. Such incentives can be a powerful companion tool to methane abatement requirements and can incentivise oil and gas companies and other market participants to invest in technologies and infrastructure to reduce methane emissions.

In order to incentivise operators to contribute to curbing emissions, governments can design **a carefully balanced reward-penalty system** with a view to ensuring that it is cost-effective for companies to deploy technologies and solutions that reduce methane emissions. Alongside financial rewards for oil and gas companies that take action to reduce their emissions, the system should also penalise those that do not comply with methane abatement regulations.

The reward component

The reward component can take different forms, including through direct financial support for methane abatement, cost-recovery provisions under production sharing agreements (PSAs), and market-based mechanisms.

Direct public financial support for methane abatement

Some governments provide direct financial support to oil and gas companies to reduce methane emissions from their oil and gas operations. For example, in the United States, the Inflation Reduction Act 2022 established the Methane Emissions Reduction Program which makes available USD 1.55 billion for financial and technical assistance for methane abatement in the oil and gas sector.[1] In Canada, the *Emissions Reduction Fund 2020* provides a mechanism where oil and gas companies can receive financial support to reduce routine venting of natural gas from oil and gas operations beyond compliance with regulatory requirements – see Box 8.2.

Box 8.2. Incentivising methane reductions in Canada: The Methane Emissions Reduction Program

Launched in 2020, the Emissions Reduction Fund (ERF) aims to incentivise onshore and offshore oil and gas companies invest in clean solutions to reduce GHG emissions and retain jobs in the sector. Applications to the ERF can be made by Canadian upstream oil and gas companies as well as midstream gathering and processing infrastructure companies.

Project requirements

ERF onshore projects considered for funding must eliminate routine intentional venting or flaring of methane emissions and surpass the applicable regulatory requirements for the facility/operation and must result in net emissions reductions that are verifiably incremental to what is required under the relevant regulation(s). Potential projects include, but are not limited to, the following examples:

- Projects that eliminate venting and/or flaring sources at surface facilities (e.g. single wells, multi-well batteries, gas processing, tanks, etc.)
- Pipeline infrastructure projects that will facilitate the conservation of otherwise vented and/or flared gas streams
- Projects that conserve otherwise vented or flared methane rich gas for onsite fuel use; and
- Projects that eliminate venting from pneumatic devices.

Funding requirements

In order to attract substantial projects that will have a noticeable impact, the ERF onshore programme offers a minimum of CAD 100 000 and a maximum of CAD 50 million per company. Successful applicants will be subject to the following requirements:

- up to 75% of total project cost with the option to stack with funding up to 90% from other sources, such as provincial programmes
- repayable and partially repayable contributions
- a five-year payback period after project completion
- expenses must be related to the project, such as Baseline Opportunity Assessment, salaries and benefits related to the project, capital expenses, and equipment rental.

Source: (Natural Resources Canada, 2020[2]).

However, public financing does not represent a feasible option for oil and gas developing producing countries as most of them are already fiscally constrained and highly indebted. In addition, governments would find it difficult to justify the subsidisation of methane abatement measures, given the pressing need to deliver on other development priorities, unless tangible benefits in terms of improved energy access, pollution reduction and improved public health can be demonstrated.

Making costs of compliance with methane abatement requirements cost-recoverable

The costs associated with the deployment of methane abatement technologies and compliance with MRV requirements can increase the operating costs of oil and gas projects and make them less profitable. In line with the recommendations contained in Guiding Principle VII of the 2020 OECD Development Centre's *Guiding Principles for Durable Extractive Contracts*, when changes in law entail costs of compliance, these costs should be treated as any other operational costs for the purposes of tax deductibility. In oil and gas production sharing agreements, these additional costs would be recoverable from the allocation of "cost oil" or "cost gas" (OECD, 2020[3]).

In jurisdictions where responsive fiscal terms are contemplated in contracts and/or law, lower project profitability resulting from increased costs of compliance with regulations on methane emissions abatement would automatically result in an equitable sharing of the financial benefits between the government and the investor. Governments should note that the application of cost-recovery measures may reduce governments revenue as the absorption of the costs of methane reduction technology cost may reduce the "profit oil" share for the state. This may be of particular concern for governments that take their allocation of profit oil "in kind" to meet domestic energy requirements. However, a lower share of profit oil could be compensated by volumes of captured associated gas, which could then be used for domestic power generation or other domestic needs, thus contributing to enhanced energy security. Furthermore, there is growing evidence from North America that regulations can lead to methane emissions reduction without significantly impacting production – see *Introduction*.

Treating the costs of compliance with methane abatement requirements as cost-recoverable can create value for the country when the methane can be captured and used productively. However, an equitable sharing of such costs also requires that, wherever necessary, financial support for investments in midstream and downstream segments of the gas value to capture, transport and process associated gas is available, with due consideration given to risks of asset stranding, lock-in and delayed transition plans – see *Building block: Financing methane emissions abatement*. In addition, many developing oil and gas producing countries are already largely indebted or fiscally constrained and may not prioritise this expenditure beyond what is required by the IOCs, whereas the availability of finance could lead to improved sustainable development outcomes when providing the means of implementation for LT-LEDS supporting broader systemic transformation toward net-zero energy systems.

Market-based mechanisms for methane abatement

Market-based incentives for methane abatement may include mechanisms that seek to attract third-party investors to participate in methane reduction, for example through clustering opportunities for flaring reduction and the allocation of rights to third parties to monetise associated gas at specific flaring sites through a competitive bidding process.

Market-based mechanisms can also induce companies to comply with methane abatement requirements. For example, in Canada, the Alberta Emission Offset System (AEOS) allows companies to generate credits if they can demonstrate emissions reduction beyond the targets set out in regulations, which can then be sold on the open market. Since its induction, operators have replaced existing pneumatic equipment with technology to reduce methane emission in 560 projects and have captured or reduced vented gas in 230 projects. Collectively these projects have avoided around 9 Mt CO_2-equivalent methane emissions (IEA, 2023[4]).

Enabling and incentivising the capture and utilisation of associated gas

Box 8.3. Enablers and incentives for the capture and utilisation of associated gas

Enablers

- Clear ownership allocation of associated gas and understanding of implications associated with different ownership structures
- Accurate MMRV to quantify volumes of associated gas at stake
- Required gas transmission infrastructure in place.

Incentives

- Capital expenditures borne by the operator for the storage and delivery of gas are cost-recoverable
- Fiscal incentives to encourage methane abatement
- Associated gas clustering opportunities based on transparent information on flaring profiles
- Preferential access for associated gas into the gas network
- International schemes signalling demand for methane abated associated gas.

Source: Based on (OECD, 2022[1]).

Clear ownership allocation of associated gas and understanding of implications associated with different ownership structures

Turning associated gas into an asset, rather than an unwanted by-product of oil production, requires that regulations, licences and production sharing contracts clearly determine who is entitled to use this resource. In addition, different ownership structures provide varying degrees of incentives for oil and gas companies or third-party investors to capture associated gas. There are two standard approaches to allocating ownership of associated gas.

Ownership of associated gas is vested in the operator. In this model, the operator that is producing oil from the underlying oil field also has the rights to extract, sell and utilise any associated gas. This option provides the operator with a direct financial incentive to capture the associated gas (the right to sell or use on-site). In addition, operators are well placed to monetise a deposit of associated gas due to their technical expertise and proximity to the resource. For example, in Angola, the model production-sharing agreement of the NOC Sonangol grants the operator the right to use any associated gas produced in their oil activities and separate any liquids from it. All capital expenditures borne by the operators for the storage and delivery of surplus gas to Sonangol are cost-recoverable against oil revenues (World Bank, 2022[5]).

Ownership of associated gas is reserved to the state. In this model, the operator has the right to produce oil only, and the right to capture and use any associated gas is reserved to the state. Governments should note that this option is unlikely to incentivise an operator to take steps to capture any associated gas. In jurisdictions where the associated gas ownership is reserved to the state, governments should take proactive steps to ensure that third parties are able to secure access to these resources. For example, in 2018, Nigeria enacted the *Flare Gas (Prevention and Pollution) Regulations* to empower the government to issue permits to access flare sites and take associated gas. The Nigerian Gas Flare Commercialisation Programme (NGFCP) was developed to tackle small flaring sites that proved more difficult to monetise. For each flare site included in an auction, the relevant oil producer is required to provide the annual amounts of flare gas that it expects to have available – either for a minimum of 15 years, or the expected life of the oil field. The Nigerian Upstream Petroleum Regulatory Commission (NUPRC) allocates rights to monetise associated gas at specific flaring sites through a competitive bidding process.

This market-based solution can be used to attract third-party investors to participate in associated gas monetisation to avoid wasteful gas flaring. Although the NGFCP is aimed at third-party investors, oil companies that exploit non-associated gas may also participate in this bidding process, but only through a midstream corporate entity incorporated in Nigeria. Successful bidders will need to enter into a series of commercial agreements with both the government and the relevant operator (oil producer), after which they will be granted permits to access the gas (World Bank, 2022[6]; NUPRC, 2023[7]):

- **Gas sales agreement** – the bidder and the government enter into a gas supply agreement under which the bidder buys gas from the government at the price contained in the bid. The operator is not a party to this agreement.
- **Milestone development agreement** – the bidder agrees to a milestone development agreement where they provide a financial guarantee to the government to underpin their commitment to project milestones; and
- **Connection agreement** – the bidder and the operator enter into a connection agreement. This lays out their arrangement of infrastructure and authorises the bidder to engineer, procure, and construct the gas connection infrastructure under terms acceptable to the operator. The infrastructure is then turned over to the operator who operates and maintains it.

Addressing the infrastructure deficit

Commercial viability is a significant barrier to the utilisation of associated gas, as additional investments may be required to develop the transport infrastructure that is necessary to bring the gas to market or the energy infrastructure to put gas to productive use (e.g. gas fired power plants). Where sufficient infrastructure is not in place to capture and utilise associated gas, governments will need to consider whether gas can have a role in a country's overall energy mix – see Box 8.4. For example, governments will need to consider what mechanisms should be applied to gas projects to mitigate risks of gas lock-in, with the exact nature of requirements determined by country-level circumstances. This can include system flanking measures (i.e. requirements for investors to invest in parallel in research and development and innovation or targets for renewable expansion) as well as project level requirements for future proofing gas infrastructure (for example, gas transport or distribution networks to accommodate low carbon fuels, or gas-to-power plants to switch to low carbon fuels by a certain date, sunset clauses after which a project should be decommissioned) (see, OECD *forthcoming*, *Gas Use Decision Tree Tool and Scorecard*, OECD Development Policy Tools, OECD Publishing, Paris).

Box 8.4. Assessing gas utilisation options: Infrastructure requirements

The commercial viability of the use of associated gas often depends on infrastructure considerations. Governments will need to assess whether adequate infrastructure is in place to evacuate gas, including processing and transport facilities (pipelines or LNG platforms) and distribution networks. Governments should note infrastructure CAPEX costs tend to be far higher when gas production is offshore, given additional costs associated with bringing gas onshore, unless exported in LNG form.

Above all, when considering gas utilisation options, governments should take steps to mitigate risks of stranded assets and gas and emissions lock-in.

- **Gas for export** – the commercialisation of gas resources for export may be achieved through LNG or cross-border pipelines. Infrastructure requirements include the LNG liquefaction facility and associated pipelines between the gas production site and the liquefaction facility. Floating LNG projects can mitigate to a degree risks of stranded assets given the CAPEX requirements are lower. For example, a recent Perenco and SNH project in Cameroon was able to greatly reduce CAPEX costs by reusing and recycling parts and through the use of a small-scale and modular design.
- **Gas-to-power** – investments in gas-fired-power generation can displace more carbon emitting energy sources such as heavy fuel oil and coal. Infrastructure requirements include the construction of new gas-fired power plants or the retrofit of coal-fired power plants (where applicable). Transmission infrastructure and interconnectors will also be required.
- **Industrial use** – the use of gas in industry (for example, fertiliser, steel, and cement production) can displace more expensive and polluting fuels such as heavy fuel oil (HFO) and residual fuel oil (RFO). Infrastructure requirements include the configuration of industrial facilities to accommodate gas as well as associated pipelines to supply gas to industrial clusters.
- **Transport and residential** – the use of gas as an alternative transport fuel may help reduce emissions, particularly in shipping and long-distance heavy goods transportation. Infrastructure requirements include converting ships/trucks with diesel compression ignition engines to run on natural gas (e.g. CNG, LNG etc.) and the construction/modification of refuelling and bunkering stations. The use of gas for residential purposes will require distribution networks to transport natural gas from production sites to the end consumer. Infrastructure requirements include transmission pipelines, compressor and pressure reduction stations.

Source: (OECD, forthcoming[8]).

Encouraging associated gas clustering opportunities

The often small and geographically dispersed nature of flare and venting sites may deter investments in gas utilisation. In addition, third-party investors may not have access to those sites or to venting and flare site data to take informed investment decisions. Even when oil companies are permitted to use associated gas, the rules regarding the transport and monetisation of associated gas may not be clear (World Bank, 2022[6]). Operators may also struggle to secure off-take agreements with buyers where the volumes are too small and associated gas production is inconsistent. Production facilities/sites that flare less than 1 mmscf/d are generally very challenging to monetise and therefore, a portfolio approach may be necessary to build minimum of economies of scale, and to hedge against the uncertainty and unpredictability of flare profiles by substituting any shortfall from a specific flare with gas from another one (World Bank, 2022[6]). Consequently, governments should consider the feasibility of spreading costs across a number of market participants by requiring operators on adjacent fields to collaborate to capture associated gas. Enabling factors include the collection and provision of transparent data on venting and flaring profiles, third-party access to existing gas networks, preferential market access for associated gas to the gas network and wholesale market and cooperative trade frameworks that signal demand for associated gas.

Providing transparent data on flaring and venting profiles

The first step toward capturing associated gas through clustering opportunities is for governments to provide clear information around flaring and venting profiles (e.g. average volume, flow rate etc.) to market operators. For example, in Trinidad and Tobago, a public registry of approved MRV data is managed by the Environmental Management Authority. The registry is broad in scope but includes information regarding GHG emissions sources from the upstream oil and gas sector (MPD, 2021[9]). In Guyana, the Ministry of Natural Resources' Petroleum Management Programme regularly publishes consolidated data on gas injected, flared and used as fuel online (Ministry of Natural Resources, 2023[10]). In Alberta, the Albert Energy Regulator provides information on flaring and venting emissions to licensees and operators in order to encourage and facilitate associated gas clustering opportunities (Alberta Energy Regulator, 2022[11]).

Ensuring third-party access to the gas network

In jurisdictions with existing gas networks, government should ensure that barriers do not prevent the sale of associated gas into the domestic gas network. The presence of a local monopolist may also result in market barriers to the transportation and sale of gas. In this regard, open access rules to gas networks can foster competition and provide opportunities to market associated gas downstream. For example, in Argentina, *Law No. 26.197, 2006*, provides that midstream and downstream operators of pipelines and other transport and distribution infrastructure are required to provide open access to third parties if they have available capacity (World Bank, 2022[12]).

A similar approach is taken in Norway, where section 59 of the 1997 *Regulations to Act relating to petroleum activities* sets out the principles for access to upstream pipeline networks. The upstream gas transport infrastructure on the Norwegian continental shelf is subject to regulated third-party access. An independent system operator, Gassco, grants access to the upstream pipeline network to users on objective and non-discriminatory terms to users with a duly substantiated reasonable need for transportation and/or processing capacity. The Ministry of Petroleum and Energy (MPE) determines the tariffs for regulated access to gas infrastructure and the tariff consists of a capital element and an operating element. The capital element promotes resource management and gives the owners a reasonable investment return. The operating element is set to cover all operating costs of the system. The Norwegian regulator retains an oversight of these activities and the MPE must approve any access agreements (Norwegian Offshore Directorate, 1997[13]; Holmen Brown, 2022[14]).

Granting preferential market access for associated gas

Governments may also consider granting preferential access for associated gas into the national gas pipeline system and preferential access for electricity produced from associated gas to the wholesale market. For example, in Russia, 2012 amendments to *Federal Law No. 241-FZ,* requires owners and operators of gas transmission and distribution infrastructure to give preferential access to associated gas in the "Unified Gas Supply System". In accordance with *Federal Law No. 35-FZ on the Electric Power Industry, 2003*, electricity produced from associated gas has priority access to the wholesale market (Lorenzato et al., 2022[15]).

Capitalising on international frameworks signalling demand for associated gas

Governments should also consider capitalising on emerging specific international demand for associated gas. For example, in its 2022 *REPowerEU* toolbox, the European Commission announced the creation of "You collect/ we buy" scheme to incentivise the capture of gas that is currently wasted (through flaring, venting, or fugitive leaks) in gas producing countries (Piebalgs and Olczak, 2023[16]; IEA, 2022[17]).

The scheme will be launched at COP29 in Baku as the Methane Abatement Partnership Roadmap with an additional emphasis on partnerships between importers and exporters. The Roadmap provides a globally adaptable, step-by step blueprint for the implementation of importer-exporter partnerships, and sets out several pillars for a co-operation framework, including a robust MRV system built on the OGMP 2.0 framework, complemented by other relevant measures and policies, as well a project plan on the timeline, abatement targets, expenditure, investments and available tools in co-operation with organisations, such as the IEA, IMEO, OECD and CCAC. The Roadmap aims to mobilise efforts under the Global Methane Pledge, incentivise importer-exporter co-operation in support of companies improving their MMRV abilities to mitigate methane emissions, and attract private investments – all while contributing to both decarbonising energy systems but also ensuring security of supply.

The penalty component

To incentivise compliance, the incentives discussed above should be combined with financial charges, such as fees or taxes, to penalise companies that fail to comply with regulations on methane emissions abatement. For fees and taxes to be effective in encouraging compliance, the cost of compliance with emissions regulations should be lower than any imposed charges.

Several jurisdictions impose a financial charge or tax on flaring and venting in order to encourage cost-effective methane emission reduction. This approach is taken in Brazil, Guyana, Nigeria and Norway (see Box 8.5 below). For example, in Brazil, the regulatory agency outlines annual and monthly limits for flaring and venting. If operators exceed these limits, they are obligated to pay royalties on the methane that is unnecessarily flared or vented. In Guyana, the government introduced a specific tax on flaring in 2022 in order to dis-incentivise flaring. The amount was set at USD 45 per tonne of CO_2 before being raised to USD 50. Notably, in addition to this (environmental) tax on flaring, operators who flare gas must also make (economic) payments to the Guyanese state for their share of the gas that was flared. In 2022, Exxon paid some USD 9 million in flaring fees to Guyana's Environmental Protection Agency (Government of Guyana, 2022[18]; Kaieteur News, 2024[19]). In Nigeria, the *Flare Gas (Prevention of Waste and Pollution) Regulations 2018* imposes taxes on flared gas. Operators that produce more than 10 000 barrels of oil per day, must pay USD 2.00 for each 28.317 m^3 of gas flared. Smaller facilities pay USD 0.50 per 28.317 m^3 methane flared (IEA, 2021[20]).

Box 8.5. GHG taxation in Norway: Incentivising the deployment of methane mitigation technologies

Norway was one of the first countries to introduce an offshore carbon tax in 1990 through the adoption of the *Act 21 December 1990 no 72 relating to tax on discharge of CO_2 in the petroleum activities on the continental shelf* (hereafter "the Act"). Although the primary target of this tax was CO_2 due to the high CO_2 content in the Sleipner field, the tax also applies to methane emissions resulting from offshore oil and gas production. Sections 1, 2, and 4 of the CO_2 Tax Act, 1990, require operators to pay on behalf of all licensees a carbon dioxide tax payment for flared or vented natural gas and any other carbon dioxide discharged to the atmosphere during the production and transport of oil and gas unless otherwise exempted by the parliament. Section 3 of the Norwegian Petroleum Act, 1996 coherently allocates to the licensees the ownership of all oil and gas produced, including gas that is flared or vented. Under the Norwegian Petroleum Act, 1996, flaring in excess of what is needed for safe operations is prohibited and subject to a fine, as is the wilful or negligent submission of incorrect or incomplete documentation or any other breach of provisions or decisions contained in or issued by virtue of the CO_2 Tax Act, 1990.

Several factors were influential in the implementation of this tax, including:

- Upstream operations on the Norwegian Continental shelf account for about 95% of the total methane emissions from the oil and gas sector.
- Operators use similar equipment which facilitates procedures for consistently calculating emissions across companies. The Act requires oil and gas companies to install metering systems to obtain methane measurements for tax purposes. Direct measurements (such as flow meters on vent heads) account for around two-thirds of emissions, while operators must follow recognised quantification models and methods to compute emissions for the remaining one-third.
- Norwegian regulators held extensive consultations with industry, research institutions, and other actors during the development of guidelines for emission data collection and reporting.

Since 1991, the CO_2 tax level has been increased and extended from offshore to other onshore industry sectors. The tax rate in 2021 was NOK 8.76 per standard cubic metre of emissions of natural gas, which is equivalent to approximately USD 1 600 or EUR 1 500 per tonne of methane.

The imposition of the CO_2 tax has encouraged investment in CO_2 and methane mitigation technologies, specifically the deployment of carbon capture and storage in natural gas production. The CO_2 tax was one of the main business drivers for Equinor to separate CO_2 offshore and inject it into deeper geological layers. Due to the Norwegian CO_2 emissions tax, it was more economical to store the CO_2, once captured, than venting it. Had this process not been adopted and the CO_2 produced been allowed to escape to the atmosphere the licensees of the Sleipner West field would have had to pay around NOK 1 million/day in Norwegian CO_2 taxes.

Source: (OECD, 2022[11]); (IEA, 2021[20]); (Mohlin et al., 2022[21]); (Vernon et al., 2022[22]); (World Bank, 2024[23]).

In the United States, the *Inflation Reduction Act 2022* introduced a methane fee on methane emissions from oil and gas operations. Section 60113 establishes the *Methane Emissions Reduction Program* that introduces a charge on methane emissions by oil and gas companies who report emissions under the *Clean Air Act*. This charge applies to facilities that emit over 25 000 metric tons of CO_2 equivalent per year, and starts at USD 900 per metric ton of methane in 2024, ramping up to USD 1 500 over a three-year period. However, the application of the charge is subject to statutory exemptions, including: where there is an unreasonable delay in permitting infrastructure to capture methane releases; when wells have been plugged in accordance with applicable requirements; and where methane releases occur at equipment that is in compliance with regulatory standards. The Inflation Reduction Act also imposes a royalty obligation on all gas produced, including gas that is consumed or lost by flaring, venting, or fugitive releases during upstream operations on federal lands and waters (IEA, 2023[24]).

Governments should ensure that regulators are empowered to enforce regulations and properly resourced to monitor compliance. Regulators will need a system to receive, process and interpret large volumes of data provided by oil and gas companies that are subject to MRV requirements. In some jurisdictions, regulators may rely on third party verifiers rather than developing the requisite in-house audit resources. Third-party verifiers may carry out similar activities as government auditors, including inspections, analysis of reports or undertaking specific monitoring/measurements campaigns. In all cases, regulators will need the technical ability to detect non-compliance as well as the political authority to bring enforcement actions for non-compliance, including monetary penalties, removal of privileges or other sanctions (IEA, 2021[20]).

Notes

[1]Alongside financial rewards for oil & gas companies that take action to reduce their emissions, the Inflation Reduction Act 2022 also sets out provisions to penalise those that do not comply with methane abatement regulations – see sub-section below – "penalty component".

References

Alberta Energy Regulator (2022), *Directive 060: Upstream Petroleum Industry Flaring, Incinerating, and Venting*, Alberta Energy Regulator, https://www.aer.ca/regulating-development/rules-and-directives/directives/directive-060. [11]

Government of Guyana (2022), *Guyana's Low Carbon Development Strategy 2030*, Government of Guyana, https://lcds.gov.gy/wp-content/uploads/2022/08/Guyanas-Low-Carbon-Development-Strategy-2030.pdf. [18]

Holmen Brown, H. (2022), *Legal framework of petroleum activity on the Norwegian Continental Shelf*, Thommessen, https://www.thommessen.no/en/news/legal-framework-of-petroleum-activity-on-the-norwegian-continental-shelf. [14]

IEA (2023), *Financing reductions in oil and gas methane emissions*, International Energy Agency, Paris, https://www.iea.org/reports/financing-reductions-in-oil-and-gas-methane-emissions. [4]

IEA (2023), *Inflation Reduction Act of 2022*, International Energy Agency, https://www.iea.org/policies/16156-inflation-reduction-act-of-2022. [24]

IEA (2022), *How to Avoid Gas Shortages in the European Union in 2023*, International Energy Agency, Paris, https://www.iea.org/reports/how-to-avoid-gas-shortages-in-the-european-union-in-2023. [17]

IEA (2021), *Driving Down Methane Leaks from the Oil and Gas Industry*, International Energy Agency, Paris, https://www.iea.org/reports/driving-down-methane-leaks-from-the-oil-and-gas-industry. [20]

Kaieteur News (2024), *World Bank highlights need to reduce emissions as carbon tax payments increased in 2023*, Kaieteur News, https://www.kaieteurnewsonline.com/2024/05/25/world-bank-highlights-need-to-reduce-emissions-as-carbon-tax-payments-increased-in-2023/#:~:text=In%20Guyana%2C%20the%20government%20instituted,gas%20during%20oil%20production%20activities. [19]

Lorenzato, G. et al. (2022), *Financing Solutions to Reduce Natural Gas Flaring and Methane Emissions*, World Bank, https://www.worldbank.org/en/programs/gasflaringreduction/publication/financing-solutions-to-reduce-natural-gas-flaring-and-methane-emissions. [15]

Ministry of Natural Resources (2023), *Data Visualization*, Petroleum Management Programme, https://petroleum.gov.gy/data-visualization?tid=All&tid_1=All. [10]

Mohlin, K. et al. (2022), *Policy instrument options for addressing methane emissions from the oil and gas sector*, Environmental Defense Fund, https://papers.ssrn.com/sol3/papers.cfm?abstract_id=4136535. [21]

MPD (2021), *Trinidad and Tobago's First Biennial Update Report*, Ministry of Planning and Development, https://unfccc.int/sites/default/files/resource/FIRST_%20BUR_TRINIDAD_AND_TOBAGO.pdf. [9]

Natural Resources Canada (2020), *Emissions Reduction Fund (ERF)*, Natural Resources Canada, https://natural-resources.canada.ca/science-and-data/funding-partnerships/opportunities/current-funding-opportunities/emissions-reduction-fund/onshore-program-emissions-reduction-fund/erf-faq/23827. [2]

Norwegian Offshore Directorate (1997), "Last amended 23 December 2022", in *Regulations to Act relating to petroleum activities*, Norwegian Offshore Directorate, https://www.sodir.no/en/regulations/regulations/petroleum-activities/. [13]

NUPRC (2023), *Nigerian Gas Flare Commercialisation Programme (NGFCP)*, Nigerian Upstream Petroleum Regulatory Commission (NUPRC), https://ngfcp.nuprc.gov.ng/. [7]

OECD (2022), *Equitable Framework and Finance for Extractive-based Countries in Transition (EFFECT)*, OECD Development Policy Tools, OECD Publishing, Paris, https://doi.org/10.1787/7871c0ad-en. [1]

OECD (2020), *Guiding Principles for Durable Extractive Contracts*, OECD Development Policy Tools, OECD Publishing, Paris, https://doi.org/10.1787/55c19888-en. [3]

OECD (forthcoming), *Gas Use Decision Tree Tool (GUDTT) and Scorecard*, OECD Publishing, Paris. [8]

Piebalgs, A. and M. Olczak (2023), *Policy Brief: The EU can reduce global methane emissions by jointly purchasing gas*, Florence School of Regulation (FSR), https://cadmus.eui.eu/bitstream/handle/1814/75327/QM-AX-23-001-EN-N.pdf?sequence=1&isAllowed=y. [16]

Vernon, N. et al. (2022), *How to Cut Methane Emissions*, International Monetary Fund, https://www.elibrary.imf.org/view/journals/066/2022/008/066.2022.issue-008-en.xml. [22]

World Bank (2024), *Global Flaring and Venting Regulations Database - Norway*, World Bank, https://flaringventingregulations.worldbank.org/norway. [23]

World Bank (2022), *Financing Solutions to Reduce Natural Gas Flaring and Methane Emissions*, World Bank, http://hdl.handle.net/10986/37177. [6]

World Bank (2022), *Global Flaring and Venting Regulations: 28 Case Studies from Around the World*, World Bank, https://thedocs.worldbank.org/en/doc/fd5b55e045a373821f2e67d81e2c53b1-0400072022/related/Global-Flaring-and-Venting-Regulations-28-Case-Studies-from-Around-the-World.pdf. [12]

World Bank (2022), *Global Flaring and Venting Regulations: A Comparative Review of Policies*, World Bank, https://thedocs.worldbank.org/en/doc/fd5b55e045a373821f2e67d81e2c53b1-0400072022/global-flaring-and-venting-regulations-a-comparative-review-of-policies. [5]

9 Financing methane emissions abatement

Governments and the oil and gas industry need to take urgent steps to meet methane targets, including the agreement reached at COP28 to "accelerate and substantially reduce non-carbon-dioxide emissions globally, including in particular methane emissions by 2030". However, the cost of rolling out methane emission reduction technology and practices across the upstream oil and gas sector is significant. The IEA estimates that USD 75 billion in cumulative capital and operating expenditure is required globally over the period to 2030 to achieve sufficient reductions in oil and gas methane emissions. Furthermore, the majority of this expenditure (USD 45 billion) will be required in low- and middle-income countries, where sources of finance are likely to be more limited (IEA, 2023[1]; IEA, 2024[2]). These financing challenges may be exacerbated where countries are heavily indebted and fiscally constrained and in jurisdictions where facilities are owned and operated by smaller independent companies and NOCs. According to the IEA, approximately USD 36 billion is required to address methane emissions from NOCs who are responsible for the majority of emissions in Eurasia and the Middle East and around USD 12 billion is required for methane abatement at facilities owned by NOCs in low- and lower middle-income countries (IEA, 2023[1]).

Although the oil and gas sector has the largest abatement potential, given the possibility to achieve quick wins with existing cost-effective abatement solutions methane abatement finance appears to be disproportionately low if compared to the needs of the oil and gas sector. For example, a recent study on methane financing across multiple sectors found that the fossil fuel sector (including oil, gas and coal) received less than 1% of total abatement finance in 2021/22. According to the Climate Policy Institute, the oil and gas sector requires USD 7.9 billion in methane abatement investment per year by 2030 (de Aragão Fernandes et al., 2023[3]). However, estimates of external sources of finance targeted at reducing methane in the fossil fuel sector (including oil, gas and coal) total less than USD 1 billion (IEA, 2024[2]). This considerable funding gap is driven by several factors, including a lack of awareness of investment opportunities, a lack of viable projects, projects not reaching an investment threshold, and a hesitancy among investors on financing projects in the fossil fuel sector (Alberti and Naran, 2023[4]).

Article 2.1c of the Paris Agreement calls for "making finance flows consistent with a pathway towards low greenhouse gas (GHG) emissions and climate-resilient development" (UNFCCC, 2015[5]). The concept of "climate alignment" of investments and financing is based on this provision. However, there is a risk for climate aligned finance to create a "climate investment trap" if access to finance is precluded to oil and gas producing countries that are most in need of support to decarbonise their economies and shift to more sustainable, climate-resilient development pathways (Marcel et al., 2023[6]). To keep the collective target of limiting the average global temperature increase to 1.5°C within reach, decarbonisation measures that can bring drastic reductions in emission intensity will need to be financed across all sectors of the economy, including oil and gas projects. This means that finance for the climate transition must take a dynamic and forward-looking approaches, while avoiding static views limited to what is already sustainable today (OECD, 2022[7]).

In this respect, it is worth emphasising that there is no agreed or unique way of downscaling the global temperature goal to economic sectors, actors, or countries, which can and will decarbonise following different pathways that are nationally determined and follow different trajectories (Noels and Jachnik, 2022[8]). In addition, the notions of climate mitigation alignment and consistency not only relate to scaling up finance for activities that are already Paris-aligned, but also to financing activities and economic sectors that need to undergo and implement changes to transition towards net-zero emissions, especially in high-emitting and hard-to-abate sectors (Noels and Jachnik, 2022[8]). Lastly, GHG emissions reduction is just one amongst complementary alignment-related metrics that need to be considered to make finance consistent with climate mitigation and resilience in the context of sustainable development and efforts to eradicate poverty in line with Article 2 of the Paris Agreement (Noels and Jachnik, 2022[8]; UNFCCC SCF, 2021[9]).

Governments and industry need to explore several sources of finance in order to mobilise the funding required to drive methane reductions at the pace and scale needed to meet the goals of the Paris Agreement. Alongside industry finance, other avenues are emerging to increase financing for methane abatement. These include direct public funding, international emissions pricing schemes and sustainability-linked financing (i.e. bonds, debt for climate swaps).

Oil and gas industry financing

Since methane emissions are a negative externality of oil and gas production, oil and gas companies should carry the primary responsibility to finance methane abatement measures, given that the amount of finance required represents less than 5% of the income the industry generated in 2023 (IEA, 2024[2]; AfriCatalyst, 2023[10]). For example, in 2022, oil and gas companies earned record profits, and the industry's net income doubled to nearly USD 4 trillion. According to the IEA, just 2% of this amount would be sufficient to provide all the spending in methane emissions reduction measures across the supply chain in the IEA's Net Zero Emissions by 2050 Scenario through to 2030 (IEA, 2023[1]).

At the same time, governments should recognise that investing in methane abatement technologies and compliance with methane abatement regulations comes with higher operating costs for oil and gas companies. The establishment of methane emissions regulations will be an important factor for companies to determine the allocation of capital, in particular for NOCs that may face additional constraints given competing priorities for domestic spending, especially in low- and middle-income countries, potentially limiting the amounts of available capital to invest.

Governments may consider introducing requirements to compel oil and gas industry financing for methane abatement – including through the establishment of a compliance fund or a special purpose vehicle:

- ***Compliance funds*** – oil and gas companies pay into a dedicated fund and the proceeds of that fund are then disbursed to entities to reduce methane emissions. For example, Alberta's 2020 *Technology Innovation and Emissions Reduction Regulation* requires regulated facilities to either reduce their emissions or to purchase credits from facilities that have exceeded their reduction targets, purchase offsets from unregulated entities or pay into a compliance fund (TIER fund). In 2022, based on funding from the TIER fund, the government of Alberta launched the Industrial Transformation Challenge – an annual funding competition for technologies that can reduce GHG emissions and improve the economic competitiveness of Alberta's industrial and natural resource sectors. The 2022 round of funding committed more than CAD 60 million to 14 projects worth more than CAD 225 million (Government of Alberta, 2019[11]; ERA, 2024[12]).
- ***Special purpose vehicles*** – oil and gas companies contribute to a special purpose vehicle (SPV). The SPV conducts the due diligence, measurement and repairs, and then monetises emissions reductions through direct gas sales, by generating carbon offsets, or through a fee by the operator (IEA, 2021[13]; OECD, 2022[14]).

Oil and gas industry organisations, which themselves are funded by their members, may also offer support to developing countries and their NOCs in methane emissions abatement. Ordinarily, this support is technical in nature and may include in-person training, webinars, access to data, use of satellites etc. – see Box 9.1.

Box 9.1. Oil and gas industry organisations: Support for methane abatement

Oil and Gas Climate Initiative (OGCI)

OGCI aims to support broader oil and gas industry efforts to achieve the same ambitions as its members (collective upstream methane intensity target of well below 0.2% by 2025 and near zero methane emissions by 2030). This is achieved through leadership, monitoring and mitigation efforts, and engagement and advocacy to reduce methane emissions.

For example, OGCI's flagship satellite monitoring campaign combines the acquisition of satellite data of methane emissions over numerous countries with targeted confidential engagement with relevant operators to share the data and support mitigation of detected emissions. The campaign was previously piloted in Iraq in 2021, run in Algeria, Kazakhstan and Egypt in 2022 and has expanded to cover additional countries.

Oil & Gas Decarbonization Charter (OGDC)

OGDC provides training free of charge to OGDC signatories. This includes 1-1 training on methane emissions quantification and mitigation delivered by Carbon Limits. The training leverages material that was developed in partnership with the UNEP IMEO OGMP2.0 and includes 9 modules, ranging from less advanced (intro to methane emissions and why it matters, sources of methane emissions, etc.) to more advanced (e.g. OGMP2.0 reporting framework) and includes a module on "Financing methane emissions reduction projects".

In addition, OGDC provides regular webinars and regional roundtables for OGDC signatories to share best practices and engage in a community of experts, to accelerate methane emissions reductions.

Source: Authors' elaboration based on responses to OECD survey on financing mechanisms for methane abatement.

Multilateral financing

Multilateral financial institutions

Multilateral development banks (MDBs) may provide an important source of financing for methane abatement projects through the provision of concessional finance, the provision of technical expertise, and by partnering with public and private sector stakeholders to de-risk and catalyse private investment – see Table 9.1.

Table 9.1. Multilateral financial institutions: Oil and gas financing

Institution	Summary of policy on oil and gas project financing
African Development Bank (AfDB)	The AfDB does not support oil and gas exploration but will continue to support Africa's net zero transition during which transition natural gas is a relevant resource for the continent's industrialisation, particularly for harder to abate industrial sectors, if used in line with the 1.5°C Paris Agreement goal, respective NDCs as well as Long-term Strategies, where these are available.
Asian Development Bank (ADB)	The ADB does not support oil field exploration or oil field development projects but does support financing natural-gas-based power plants as well as safety and efficiency improvements in the transportation of oil and LNG.
Asian Infrastructure Investment Bank (AIIB)	The AIIB supports fossil-fuel–based generation using commercially available least-carbon technology as well as development, rehabilitation, and upgrading of natural gas transportation, storage, and distribution infrastructure, and control of gas leakage, to foster greater use of gas as a transition to a less carbon-intensive energy mix.
CAF Development Bank of Latin America and the Caribbean (CAF)	The CAF provides support for downstream and midstream oil and gas projects. This is aligned with CAF's strategic decision to support gas projects as an energy transition fuel.
European Bank for Reconstruction and Development (EBRD)	The EBRD combines investments with policy engagement and technical assistance to support methane abatement across a range of sectors. The EBRD does not support oil and gas upstream projects unless aligned with the goals of the Paris Agreement.
European Investment Bank (EIB)	The EIB will not support upstream oil or natural gas production, or infrastructure dedicated to oil and natural gas (networks, liquefied natural gas terminals, storage).
Inter-American Development Bank (IDB)	The IDB does not support upstream oil exploration and development projects. However, on a case-by-case basis, consideration is given to financing upstream gas infrastructure where there is a clear benefit in terms of energy access for the poor and where GHG emissions are minimised, projects are consistent with national goals on climate change, and risks of stranded assets are properly analysed.
Islamic Development Bank (IsDB)	No explicit restrictions on oil and gas financing, although upstream oil and gas is excluded from certain IsDB financing mechanisms – e.g. Green and Sustainability Sukuks issued under its 2019 Sustainable Finance Framework.
World Bank Group	In 2019, the World Bank Group ceased financing upstream oil and gas projects. However, financing for natural gas projects may be provided in exceptional circumstances where there is a clear benefit in energy access for poor countries and the project is consistent with the country's Paris Agreement commitments.

Source: (AfDB, 2012[15]); (AfDB, 2024[16]); (ADB, 2023[17]); (AIIB, 2022[18]); (CAF, 2024[19]); (EBRD, 2023[20]); (EIB, 2023[21]); (IDB, 2020[22]); (IsDB, 2018[23]); (IsDB, 2019[24]); (World Bank Group, 2017[25]); (IFC, 2019[26]).

Financing oil and gas exploration and production and related activities

In recent years, many MDBs have updated their energy policies and strategies to focus on renewable energy and decarbonisation and to ensure that lending is aligned to the objectives of the Paris Agreement. Consequently, conditions for financing fossil fuel-related projects have tightened as several MDBs have announced a shift away from funding fossil fuels – particularly in respect of coal-related investments and upstream oil and gas exploration and production. For example, the Asian Development Bank (ADB), European Bank for Reconstruction and Development (EBRD) and the European Investment Bank (EIB) have ceased financing upstream oil and gas projects. The ADB will not support any natural gas exploration or drilling activities (ADB, 2023[17]). The EBRD's 2023 *Energy Sector Strategy 2024-28* provides that the EBRD will not invest in the upstream oil or gas sectors (EBRD, 2023[20]). The EIB's 2023 *Energy Lending Policy: Supporting the Energy Transformation* notes that the EIB will not support upstream oil or natural gas production, coal mining, infrastructure dedicated to coal, oil and natural gas (networks, liquefied natural gas terminals, storage) (EIB, 2023[21]).

However, some banks still continue to provide financing for upstream oil and gas projects, albeit only under exceptional circumstances. The Inter-American Development Bank (IDB's) 2020 *Environmental and Social Policy Framework* stipulates that the IDB will not finance upstream oil exploration and development projects. Financing for upstream gas exploration and development projects is ordinarily prohibited. However, financing may be provided in exceptional circumstances where there is a clear benefit in terms of energy access, where GHG emissions are minimised, where projects are consistent with national goals on climate change, and where risks of stranded assets are properly analysed (IDB, 2020[22]). In 2019, the World Bank Group ceased financing upstream oil and gas projects, except in exceptional circumstances in the poorest countries where there is a benefit to energy access and this is consistent with countries' NDC commitments (World Bank Group, 2017[25]; IFC, 2019[26]). For example, in 2021 the International Finance Corporation (IFC) provided financing to reduce flaring, improve energy access, and to support a more resilient, sustainable energy sector in Iraq – see Box 9.2.

Box 9.2. IFC financial support for gas flaring reduction in Iraq

Iraq is a major international oil producer but also has significant reserves of natural gas produced as a byproduct of oil extraction. In 2011, Iraq joined the World Bank's Global Gas Flaring Reduction Partnership (GGFR) [since renamed as the Global Flaring and Methane Reduction Partnership (GFMR)] and in 2013, committed to eliminate all routine natural gas flaring by 2030. However, given the absence of adequate infrastructure to capture and process it, about 70% of natural gas produced in Iraq is flared.

Project details

In 2021, the IFC announced its investment in the Basrah Gas Company (BGC) to support one of the largest gas flaring reduction projects in the world. BGC is a 25-year incorporated joint venture between Iraqi NOC South Gas Company (51%), Shell (44%) and Mitsubishi Corporation (5%) that was created to capture and process associated gas that would otherwise be flared.

The project is expected to increase BGC's gas processing capacity, thereby avoiding flaring and reducing associated GHG emissions by around 10 million tons per year. The project is intended to support Iraq's transition toward a lower carbon pathway, improve access to energy to meet growing power needs, and contribute to a more resilient, sustainable energy sector in Iraq.

Details of financial structure

The IFC is the lead arranger of the five-year loan to BGC worth USD 360 million which is structured as follows:

- USD 137.76 million loan from the IFC's own account
- USD 180 million loan in which participations were syndicated to eight international banks (Bank of China, Citi, Deutsche Bank AG, Industrial Commercial Bank of China, Natixis, Sumitomo Mitsui Banking Corporation, Société Générale and Standard Chartered Bank); and
- USD 42.24 million loan through IFC's Managed Co-Lending Portfolio Program, a platform that allows institutional investors to participate in IFC's loan portfolio.

Source: (IFC, 2021[27]).

The Asian Infrastructure Investment Bank AIIB differentiates between financing for oil projects and for gas projects. While the AIIB will only support oil sector investments under exceptional circumstances to improve basic energy access, there is more flexibility for financing natural gas.

The African Development Bank (AfDB) 2012 *Energy Sector Policy* does not provide support for oil and gas exploration activities. However, the AfDB *Ten-Year Strategy (2024-2033)* acknowledges that the AfDB will continue to support Africa's net-zero transition during which transition natural gas is a relevant resource for the continent's industrialisation, particularly for harder to abate industrial sectors, if used in line with the 1.5°C Paris Agreement goal, respective NDCs as well as LT-LEDS, where these are available. At the same time, the AfDB will support the acceleration of renewable energy investments and the development of sustainable alternatives to natural gas for industrialisation and securing of related energy security (AfDB, 2024[16]). For climate change mitigation projects, the GHG emission reductions potential is one of the key indicators that the AfDB will assess during appraisal and these reductions will be measured during project implementation and supervision.

The Islamic Development Bank (IsDB)'s 2018 *Energy Sector Policy: Sustainable Energy for Empowerment and Prosperity* allows for broad financing of upstream oil and gas projects. The IsDB will examine the provision of energy forms such as oil and gas and of the relevant downstream infrastructures, based on the principles of safety, operational efficiency and sustainability (IsDB, 2018[23]). Upstream oil and gas is however, excluded from certain IsDB financing mechanisms, including Green and Sustainability Sukuks issued under its 2019 *Sustainable Finance Framework* (IsDB, 2019[24]).

Financing midstream and downstream natural gas infrastructure

Despite the shift away from financing oil and gas exploration and production, some MDBs continue to provide support for midstream and natural gas infrastructure. For example, the Development Bank of Latin America and the Caribbean (CAF) provides support for downstream and midstream oil and gas projects. This is aligned with CAF's strategic decision to support gas projects as an energy transition fuel. For example, in 2023, CAF approved a loan of USD 540 million for the North Gas Pipeline Reversion Project to construct 122.5 km of gas pipeline to allow the transportation of natural gas to the provinces in northern and central Argentina. Regional integration is another area of CAF focus. For example, in 2023 CAF financed a study to promote energy integration between Argentina and Chile, focusing on the increase of gas exports (CAF, 2024[19]).

The ADB, the AIIB and the EBRD have developed methodologies to assess whether investments in fossil fuel-based infrastructure are aligned with the Paris Agreement temperature goals. This means avoiding financing investments in emission reduction opportunities that have the effect of locking in emissions. Such investments slow down the adoption of net-zero alternatives, and result in assets needing to be replaced before the end of their lifetime, when net-zero alternatives become commercially available (OECD, 2022[6]).

For example, the ADB recognises the role of natural gas as a transitional fuel, and therefore may finance investments in natural gas transmission and distribution pipelines, LNG terminals, and storage facilities subject to a set of screening criteria consistent with the Paris Agreement:

i. No other low-carbon or zero-carbon technology, or combination thereof, can provide the same service at an equivalent or lower cost at a comparable scale.
ii. The project's operating lifetime is consistent with the carbon stabilisation trajectory aiming to achieve carbon neutrality by about 2050, and by a time set by developing member countries that are consistent with their NDCs. The project also avoids long term lock-in into carbon infrastructure and the associated risk of creating stranded assets.
iii. The project is economically viable considering the social cost of carbon and an operating lifetime consistent with condition (ii).

In addition, the ADB may support natural gas-based power generation that employs high efficiency and internationally best available technologies and reduces emissions by directly displacing other fossil-fuel-based thermal power capacity (ADB, 2023[17]).

The AIIB will provide support for mid-stream gas infrastructure (LNG terminals, storage, and transmission pipelines), natural gas-fired power generation, and downstream (distribution and end-use) facilities (AIIB, 2022[18]).

The AIIB's 2022 *Energy Sector Strategy: Sustainable Energy for Tomorrow* acknowledges that natural gas will play a transitional role in the energy system of many developing countries. Financing for mid-stream gas infrastructure, natural gas-fired power generation, and downstream facilities is available – subject to the following criteria:

- investments do not conflict with a country's climate policy and commitments including its NDC, long-term low GHG emissions development strategies and net-zero/carbon neutrality pledges
- investments do not create a risk for carbon lock-in or stranded assets – taking into account a long-term decarbonisation trajectory of the country that is consistent with the mitigation goals of the Paris Agreement
- investments reduce the energy sector's carbon intensity. Appropriate project goals may include, replacing higher carbon fuels, inefficient technologies, oil- and coal-fired energy facilities, or supporting the integration of renewable energy
- investments utilise best available technologies and sector best practices in limiting methane emissions (AIIB, 2022[18]).

The EBRD also sets out criteria where, in exceptional cases, targeted support for fossil-fuel investments in the mid- and downstream oil and gas sectors could be possible. The EBRD's 2023 *Energy Sector Strategy 2024-28* provides that such investments must not only be aligned with the goals of the Paris Agreement, but go beyond that requirement to demonstrate strong ambition to accelerate the low-carbon transition. Specifically, these projects must demonstrate:

- alignment with the goals of the Paris Agreement as per the EBRD's Paris Agreement alignment methodology
- consistency with NDCs and long-term low GHG emissions development strategies
- a low risk of carbon lock-in and therefore do not lead to carbon lock-in
- for projects with significant emissions, be subject to an economic viability test incorporating a shadow carbon price
- be located in a policy context that demonstrates commitment to the goals of the Paris Agreement, and be consistent with this policy context
- transition with a credible low-carbon pathway, either at a national or sectoral level, as per the EBRD's Paris Agreement alignment methodology
- they do not displace renewable sources or low emissions alternatives
- consistency with the Bank's Environmental and Social Policy (including requirements for using best available techniques); and
- that they would not lead to stranded assets and therefore be subject to a thorough assessment of climate-financial risks (EBRD, 2023[20]).

The criteria used by ADB, AIIB and EBRD for climate alignment assessments provide examples of methodologies that attempt to take into account geographical specificity, diversity of pathways to achieve global emissions reduction goals, and their applicability in developing countries reflecting real-economy considerations. Fostering progressive convergence and common approaches to assess eligible investments compatible with the Paris Agreement based on climate and complementary alignment-related metrics could facilitate international investment in midstream and downstream gas infrastructure, without unnecessarily constraining national decision-making on how the low-carbon, sustainable transition will be undertaken (OECD, 2020[28]).

Technical support for methane abatement in the oil and gas sector

Alongside financing, several MDBs also offer technical support for methane abatement in the oil and gas sector. For example, the AIIB has supported initiatives to encourage the adoption of national methane abatement policies and regulatory frameworks and has supported collaborative efforts to reduce routine gas flaring and fugitive methane emissions (Alberti and Naran, 2023[4]). The EBRD has recognised the central importance of abating fugitive methane emissions as rapidly as possible and has pledged to engage with its countries of operation to address this issue (EBRD, 2023[20]). For example, the EBRD has continued to offer grants directly to governments, including Kazakhstan and Uzbekistan, to develop methane emissions reduction programmes (IEA, 2023[1]). Lastly, the World Bank Group, through its subsidiaries, provides the following methane abatement support:

- The IFC provides financing and advisory services to projects that reduce gas flaring or fugitive methane emissions in existing oil and gas installations – see Box 9.2 above. Projects funded by the IFC directly or indirectly (through financial intermediaries or investment in third-party funds) are subject to ex ante and ex post verification (IFC, 2017[29]).
- The Global Flaring and Methane Reduction Partnership (GFMR), a multi-donor trust fund composed of governments, oil companies, and multilateral organisations, provides technical and regulatory support to reduce flaring and methane emissions (World Bank, 2022[30]) – see Box 9.3.

Box 9.3. The GFMR: Providing financial and technical support for methane abatement projects in developing countries

The World Bank's Global Flaring & Methane Reduction (GFMR) trust fund provides governments and state-owned operators in developing countries with technical and financial support for methane abatement projects along the entire oil and gas value chain. Specifically, to:

- conduct campaigns to detect, measure, monitor, and report methane emissions
- adopt best practices in policy and regulations, targeting low-emission energy supply practices
- unlock finance with support of financial advisory services, including a focus on financial due diligence, modelling, optimal capital structure, financial risk analysis and de-risking solutions such as guarantees, and identifying sources of commercial and concessional finance
- adopt low-emission infrastructure design and operational and maintenance practices
- assess, prioritise, identify, and select investment projects to reduce gas flaring and methane emissions; and
- implement flaring and methane emission abatement projects.

Types of support available

- **Financial support** – this takes the form of grants covering a portion of project capex – generally 10-20% of the project total cost, but this may be increased in specific contexts – for example: FCV (fragile, conflict, violence) countries, countries with very limited capacities, or countries with great difficulties accessing financial markets. Project financing may include grant funding for flaring reduction projects, methane detection and abatement projects, and LDAR programmes implemented by governments or state-owned operators, with close supervision by the World Bank.
- **Technical support** – this is delivered through analytics and studies funded by GFMR, covering all relevant aspects of a methane and flare abatement project, including technical design, regulatory aspects, emission reduction potential, and financial structure.

Criteria to assess requests for support for methane abatement measures

GFMR financial support decisions are guided by a set of pre-agreed eligibility criteria and fund allocation priorities:

- **Methane over carbon dioxide** – priority of methane emissions over gas flaring gas, recognising that methane is a much more potent GHG than carbon dioxide.
- **High volume** – the potential volume of flare gas or methane emissions that could be abated.
- **Unit cost** – opportunities will be ranked according to the estimated volume of GHG abated per USD.
- **Technical assistance needed** – priority to government entities and state-owned operators with fundamental and basic gaps in capacity to address flaring and methane emissions.
- **Government willingness and commitment** – established eligibility criteria:
    - Operator commitment to measure and report emissions accurately through OGMP2.0 or another similar comprehensive, measurement-based international reporting framework
    - Operator commitment to achieve near-zero absolute methane emissions by 2030 by reducing methane intensity to below 0.2% of the volume of the total gas marketed; and
    - Operator commitment to achieve Zero Routine Flaring by 2030 (ZRF).
- **Sustainability and replicability** – priority to projects that are part of a multi-phased, programmatic approach, or may be embedded in an existing World Bank energy sector programme in the country.
- **Funding needed** – priority to government entities and operators in fragile and conflict-affected countries or those with poor finances.

Source: Authors' elaboration based on responses to OECD survey on financing mechanisms for methane abatement.

International initiatives that provide support for methane abatement

In response to the increasing global recognition of the urgency of tackling fossil methane emissions, several international initiatives have been launched in order to drive collective global action on tackling oil and gas sectors methane emissions. Several of these initiatives provide technical support for methane abatement activities (improved data transparency, MMRV, methane abatement policies and regulations), that developing countries can access to address upstream oil and gas sector methane emissions.

The Climate and Clean Air Coalition (CCAC) is a voluntary partnership of over 160 governments, intergovernmental organisations, and non-governmental organisations founded in 2012, and hosted by UNEP. The CCAC offers funding and technical support for signatories of the Global Methane Pledge to develop national methane roadmaps or action plans – see Box 9.4. Specifically:

- **Fossil Fuel Regulatory Programme (FFRP)** – CCAC provides financial support for project funding through the FFRP for policy and regulatory development, capacity building, and technical assistance in the fossil fuel sector to reduce short-lived climate pollutants including methane and black carbon. From mid-2024 to mid-2027 the FFRP will support up to 20 developing country governments with tailored support for capacity development, regulatory frameworks, and enforce compliance with existing frameworks.
- **Additional project funding** – the CCAC Trust Fund provides financing for developing countries to enable actions that deliver climate, air quality and development benefits. Project funding is split between national policy development and implementation of mitigation measures. Recent methane abatement funding projects include the delivery of a methane emissions inventory, mitigation plan

and MRV framework for the oil and gas sector in Gabon and the delivery of training workshops and guidance for methane leak detection and repair in the oil and gas sector in Côte d'Ivoire.

- **Technical assistance** – CCAC can support countries eligible for official development assistance by providing access to technical experts within our partnership and small-scale grants for short-term services, such as guidance on technology options, funding opportunities, and policy development. Technical assistance services typically last less than 12 months and are small in scale (up to USD 50 000). In particular, CCAC supports developing countries to develop methane roadmaps and design policies to reduce methane emissions to realise the goals of the GMP. These services may include regulatory analysis, cost-benefit analysis, and peer-to-peer exchanges.

Box 9.4. CCAC funded projects: Supporting methane reduction in Gabon and Côte d'Ivoire

Gabon – Methane emissions inventory, mitigation plan and MRV framework for the oil and gas sector

Gabon is the fifth-largest oil producer in Sub-Saharan Africa. While the country also has considerable associated gas resources, more than 90% of gas production is either re-injected or flared for lack of economic alternatives. In addition, Gabon's flaring intensity (i.e. gas flared per unit of oil produced) is also significant – representing the second highest jurisdiction after the Bolivarian Republic of Venezuela in a recent study by the World Bank.

Gabon has recognised the importance of tackling methane emissions from the upstream oil and gas sector. In 2015, Gabon submitted its NDCs to the UNFCCC, highlighting its commitment to reducing greenhouse gas emissions by at least 50% by 2025. Gabon has also joined the GMP.

To meet these targets, Gabon's General Direction of Environment and Natural Protection at the National Climate Council engaged with CCAC to deliver an improved and updated methane emission inventory, mitigation assessment and MRV framework for the oil and gas sector.

Project details

Between 2023 and 2025, CCAC, through its implementing partners, CATF, will carry out the following activities to support Gabon's commitment to reduce methane emissions in the upstream oil and gas sector:

- Develop a methane emissions inventory for the Gabonese oil and gas sector using CoMAT
- Develop a methane mitigation plan and an MRV framework for the oil and gas sector
- Map currently conducted and ongoing studies of relevant activities and policies within the oil and gas sector; and
- Draft a report on gaps and recommendations for utilisation of methane from the oil and gas sector.

Côte d'Ivoire – Training workshops and guidance for LDAR in the oil and gas sector

Following recent discoveries, Côte d'Ivoire has aspirations of becoming a major African oil and gas producer. In order to ensure that new and existing gas infrastructure does not contribute to increase methane emissions, policy makers and regulators in Côte d'Ivoire will need a comprehensive understanding of modern LDAR methods and related best practices and regulatory requirements.

In this regard, Côte d'Ivoire's Ministry of Environment and Sustainable Development engaged with CCAC to develop training on the design and implementation of LDAR programmes at oil and natural gas facilities, as well as provide a review or related regulatory requirements in other jurisdictions.

Project details

Between 2023 and 2024, CCAC, will carry out the following activities to support Côte d'Ivoire's capacity to develop and implement methane leak detection and repair protocols. This includes:

- training on the key objectives of an LDAR programme, the characteristics of fugitive equipment leaks, designing and implementing an LDAR programme, and LDAR equipment selection
- catalysing actions to manage fugitive equipment leaks in Côte d'Ivoire's oil and gas sector; and
- the creation of relevant benchmarks against which government and industry can assess the effectiveness of their related actions.

Source: (CCAC, 2023[31]); (GFMR, 2023[32]); (CCAC, 2023[33]).

The European Commission identified methane emissions as an important and urgent issue requiring action in the European Green Deal in 2020. The European Commission's objective is to reduce methane emissions in the EU but also to support similar action internationally. In that regard, the Commission provides support that developing countries can draw on for assistance in reducing methane emissions.

The Commission uses criteria to assess requests for support for methane abatement measures in upstream oil and gas projects. Activities are only financed if they are clearly consistent with the overarching goal of transitioning away from fossil fuels. In addition, requests must not contravene the exclusion criteria listed in Article 29 of the Regulation (EU) 2021/947, namely, funding cannot support actions or measures which:

- may result in the violation of human rights in partner countries
- are incompatible with the recipient country's NDCs
- promote investments in fossil fuels
- cause significant adverse effects on the environment or the climate (unless such actions or measures are strictly necessary and are accompanied with appropriate measures to avoid, prevent or reduce and, if possible, off-set these effects).

While the Commission's funding is often channelled through implementing partners, notably UNEP's CCAC and IMEO, direct funding can be provided for country-specific projects (e.g. in Turkmenistan). Developing countries wishing to access Commission funding can engage with the Directorate-General for Energy's (DG ENER) focal points, who are the lead Directorate for the European Commission's engagement as champion of the Global Methane Pledge, as well as with the Directorate-General for International Partnerships (INTPA), the Directorate-General for Neighbourhood and Enlargement Negotiations (DG NEAR) and EU delegations.

Strategic investment funds

Strategic investment funds (SIFs) may also form an important source of financing for methane abatement. SIFs are special purpose investment vehicles, backed by governments or other public institutions, that pursue a dual mandate of both financial returns and policy objectives, and aim to mobilise commercial capital for investments where private investors are not available.[1] Investment strategies of SIFs are driven by their policy objectives. These objectives will vary among different SIFs but can include accelerating a country's economic development, infrastructure development, employment creation, and climate change mitigation and adaptation (World Bank, 2022[30]).

SIFs are likely to have flexibility to provide financing for methane abatement. Research by the World Bank found that 20 SIFs included oil and gas within their investment mandate (although only 16 of those SIFs appeared to have made investments in the oil and gas sector). SIFs domiciled in oil and gas economies are likely to have even more flexibility to fund projects in the oil and gas sector and therefore may be able to provide financing for methane abatement. For example, the Nigeria Sovereign Investment Authority provided USD 650 million to the Nigeria Infrastructure Fund (NIF). The NIF fund has a broad investment mandate which includes the ability to invest in gas pipelines, infrastructure, and storage projects, alongside the broad power sector (World Bank, 2022[30]).

There are several SIFs in the Middle East that actively invest in the oil and gas sector and that may be able to provide financing for methane abatement. SIFs have a strong understanding of their country's investment environment, including access to its project pipeline, and often play the role of project developers as well as financier (World Bank, 2022[30]). In the United Arab Emirates, the Mubadala Investment Company PJSC owns 100% of Mubadala Energy – an international energy company, headquartered in Abu Dhabi, with activities in 11 countries. Mubadala Energy's portfolio is 66% natural gas and views gas as a key bridging fuel to a lower carbon future (Mubadala Energy, 2024[34]). In another example, the Kuwait Investment Authority (KIA) holds a 15% share in the Arab Petroleum Pipelines Company – an Egyptian oil and gas service provider that owns and operates pipelines and associated storage facilities connecting the Red Sea and the Mediterranean. Other shareholders include: the Egyptian General Petroleum Corporation (EGPC), Mubadala Energy, Saudi Aramco and Qatar Petroleum (Mubadala Energy, 2024[35]). The newly established Africa Energy Bank can also support methane abatement projects in Africa.

Table 9.2. Strategic investment funds: Oil and gas financing

Fund	Geographic focus	Description
Africa50	Africa	Africa50 is an infrastructure investment platform founded by the AfDB and African states.
Emerging Africa Infrastructure Fund	Sub-Saharan Africa	The EAIF was launched to raise and deploy blended finance for transformative infrastructure projects across sub-Saharan Africa. Anchor investors include: the UK, Dutch, Swedish and Swiss Governments
Infraco Asia	Asia (South and Southeast)	Sectors include: power and energy; Oil and gas distribution. Note that InfraCo Asia does not participate in project development that could have damaging environmental or social impact.
Investment Corporation of Dubai	Dubai / global	ICD is the principal investment arm of the Government of Dubai that invests across a range of sectors, including upstream oil and gas.
Mubadala Investment Company	Abu Dhabi / global	MIC is an Emirati sovereign wealth fund that invests across a range of sectors, including upstream oil and gas value chain.
Nigerian Sovereign Investment Authority (National Infrastructure Fund)	Nigeria	The National Infrastructure Fund focuses entirely on domestic investments in selected infrastructure sectors, including motorways, healthcare, power, and agriculture.
Public Investment Fund	Saudi Arabia / global	The PIF is a Saudi Arabian sovereign wealth fund formed to invest in Saudi Arabia and globally, in line with Saudi Vision 2030.
Russian Direct Investment Fund	Russian Federation / global	RDIF is a Russian sovereign wealth fund, formed to co-invest alongside other investors primarily in Russia.
Silk Road Fund	China / global	The Fund is a Chinese medium- and long-term equity investment fund formed to foster increased investment in countries along the Belt and Road Initiative.

Source: (World Bank, 2022[30]).

Emissions trading schemes

Emissions trading schemes (ETS) can be powerful tools to encourage and finance methane technological innovation, deployment and scale-up as they provide long-term price signals to incentivise companies to reduce their methane footprint. Across industrialised economies, ETS tend to be the most effective mechanisms to reduce emissions as they ensure environmental effectiveness and incentivise use of the most efficient technologies and those that cost least. Therefore, as part of a long-term strategy, fossil fuel-producer emerging and developing economies may consider how to leverage explicit carbon pricing through the introduction of a market mechanism on methane abatement such as an ETS. This could create opportunities to mobilise climate finance for methane abatement (OECD, 2022[14]).

In fact, there is already precedence for a ETS financing methane abatement projects in the oil and gas sector. The Clean Development Mechanism (CDM) established under the Kyoto Protocol allowed a country with an emissions reduction commitment to implement a mitigation project in a developing country in exchange for a UN-issued Carbon Emissions Reductions (CERs), or carbon credits, which could be used to contribute to its emissions reduction commitments or to be sold. Fossil fuel producer emerging and developing economies are particularly well-placed to generate climate finance through the use of an ETS given the many methane emissions abatement opportunities across their oil and gas value chains. For example, oil and gas methane projects can create these credits through the capture and utilisation of associated gas or by implementing LDAR in natural gas facilities. Since inception of the CDM, there has been 45 oil and gas methane projects – including in Bangladesh, India and Oman. Although, governments should note that the future of the CDM is uncertain pending the outcome of negotiations under Article 6 of the Paris Agreement governing the functioning of an international carbon market (IEA, 2023[1]; OECD, 2022[14]).

Box 9.5. Funding methane mitigation in Bangladesh: The Clean Development Mechanism

Natural gas is a significant source of primary energy in Bangladesh for both industrial and residential purposes. Natural gas is distributed nationally through an aging gas pipeline network that contributes to Bangladesh's total methane emissions through leaks.

To address this challenge, the CDM has been utilised to fund methane abatements projects. Under the scheme, gas distribution companies engage third parties to identify and repair methane leaks to improve operational efficiency and safety (e.g. LDAR activities). For example, in 2021, Titas Gas Transmission and Distribution Company Limited (Titus), a gas supplier, signed a Certified Emission Reductions Project Investment Agreement with Danish company NE Climate A/S (NES) to deploy LDAR activities at Titus infrastructure. This project was registered by the UNFCCC in 2025 and has reduced methane emissions by around 4 million metric tons of CO_2 equivalent annually.

Source: (CLDP, 2023[36])

On the basis of the model established by the CDM, a number of regional economic groupings and in some cases provincial governments, have set up their own ETS – see examples from Alberta in Box 9.6. As of 2021, there are 33 ETS schemes in operation around the world covering 16% of global GHG emissions and jurisdictions making up 54% of global GDP. Furthermore, sixty-one countries, including major fossil fuels producer countries such as Egypt, Kazakhstan and Nigeria, have signalled their intention to utilise carbon markets to meet their NDC commitments (OECD, 2022[14]).

However, governments should be cognisant of the risks of the CDM (or similar schemes) creating perverse incentives that may encourage the continued and expanded use of fossil fuels. For example, this may include incentivising oil and gas producers to dilute methane in order to continue trading in emission reductions, where abatement above a certain threshold is required by regulation or advancing an investment in methane abatement that would have happened anyway (OECD Development Centre, 2020[37]). To avoid this, ETS should set incremental performance-based methane abatement requirements.

Box 9.6. Financing methane emissions reductions through emissions trading schemes: The Alberta Emission Offset System

In 2015, the Government of Alberta set a target to achieve a 45% reduction in oil and gas methane emissions by 2025 (relative to 2014 levels). In pursuit of this target, Alberta is using a combination of regulatory requirements and economic instruments to create incentives for companies to reduce emissions. The *Technology Innovation and Emissions Reduction Regulation* sets out a mechanism to reduce emissions at facilities which emit more than 100 000 tonnes of CO_2 per year, covering about 60% of Alberta's emissions. To meet this emissions reduction requirement, facilities can either: reduce their emissions; submit emission performance credits; submit emission offset credits; or pay into a compliance fund.

If companies can demonstrate that they have reduced their emissions beyond the targets set out in regulations, they can generate credits under the Alberta Emission Offset System (AEOS), which can then be sold on the open market. Companies participating in the AEOS must be registered in the Alberta Emission Offset Registry and must undergo a third-party verification process to determine that emissions have been measured in accordance with Alberta's quantification protocols. Since its induction, operators have replaced existing pneumatic equipment with technology to reduce methane emission in 560 projects and have captured or reduced vented gas in 230 projects. Collectively these projects have avoided around 9 Mt CO_2-equivalent methane emissions.

The AEOS encourages third party companies to provide financing for emissions reductions without imposing any direct costs on the asset owner or operator. Under this model, third party companies provide finance for emissions reductions at specific sites/facilities, and recoup their costs through selling emissions credits that have been earned by the project and from selling captured methane that would have otherwise been lost. Once the upfront capital expenditure has been recuperated or the project breaks even, proceeds are shared with other project partners, including the asset owner. Since 2017, around 200 methane reduction projects have been developed for 35 companies, saving an estimated 1.7 Mt CO_2-eq of methane emissions.

The AEOS has created an opportunity for third party service providers to assist oil and gas operators to finance the deployment of emissions reduction measures that otherwise would not have had the necessary financial capacity. The IEA has noted how this approach may be of value to companies with limited investment capacity, including NOCs and companies in low- and middle-income countries.

Source: (Government of Alberta, 2019[11]); (IEA, 2023[1]).

Sustainability-linked financing

Green, transition, and sustainability-linked bonds are linked to the sustainability performance of a company or by compliance with external third-party criteria and represent an important mechanism through which governments, multilateral institutions and the private sector can raise finance in a range of low-carbon investments, including methane abatement projects in the upstream oil and gas sector.

Green bonds are fixed-income debt securities which offer investors relatively low-risk returns over a given period of time. The first green bond was issued by the EIB in 2008. Since then, the green bond market has grown quickly, and accelerated rapidly following the signing of the Paris Agreement in 2015. Green bond issuances in developing and emerging countries have increased significantly in recent years, with 25 countries having issued green bonds since 2012. In 2019, total issuances in emerging markets amounted to USD 52 billion. For example, in 2017, Repsol, the Spanish energy company, issued a green bond for EUR 500 million with a five-year maturity. The proceeds were linked to energy efficiency projects and low-emission technologies, including reductions in flaring and methane emissions mitigation, and sought to avoid around 1.2 Mt CO_2-eq emissions. Following maturity of the bond, Repsol produced a report setting out how the proceeds were used to achieve the green bond objectives, and these findings were independently verified by a third-party (IEA, 2023[1]). However, generally accepted green bond standards are likely to focus on purely "green objectives" and exclude methane abatement projects from the eligible uses of proceeds. Therefore, green bonds may only form a limited avenue for methane emission reduction financing (OECD, 2022[14]; World Bank, 2022[30]).

Transition bonds are a very recent initiative, and the market is still in its infancy with less than 20 issuances explicitly labelled as such, and mostly issued by non-financial corporates in Asia. Transition finance focuses on the dynamic process of becoming sustainable, rather than providing a point-in-time assessment of what is already sustainable (Cordonnier and Saygin, 2023[38]; OECD, 2022[6]). As such, transition finance does not necessarily require countries or companies to have achieved certain performance standards to be eligible for financing, but instead provides finance for countries and companies that set themselves on an ambitious and verifiable path of transition, including performance milestones and targets to be met over a certain period, measured by pre-defined and verifiable KPIs and metrics which can help balance inclusiveness with climate integrity and avoid emissions lock-in. Although transition bond standards are still in the early stages of development, the rationale for these products (i.e. to assist developing countries and companies to decarbonise) is consistent with the climate mitigation goal of reducing methane emissions in the oil and gas sector (OECD, 2022[14]; World Bank, 2022[30]).

In recent years, there has been a growing market in sustainability-linked bonds. These differ from other sustainable bonds, such as green bonds or social bonds, in that proceeds are not used exclusively to fund specific green or social projects. Instead, sustainability-linked bonds are sovereign or corporate performance-based instruments that allow governments or companies to raise finance for general purposes, while setting out sustainability performance targets that need to be achieved by the issuer. The bond's finance terms are linked to these targets and vary depending on whether the issuer achieved the predetermined target. Targets can generally cover several sustainability-related dimensions, including climate, environmental and social elements, though nearly 60% of issuances in Q1 2022 specifically targeted GHG or carbon emission reductions (OECD, 2022[14]; World Bank, 2022[30]). See Box 9.7 for the example of Eni.

Box 9.7. Structuring sustainability-linked bonds to reduce emissions in the upstream oil and gas sector – the case of Eni

In 2023, the Italian oil and gas company, Eni issued a sustainability-linked bond for EUR 1 billion with a 7-year maturity. The bond was issued for general corporate purposes but was linked to ENI's achievement of two specific Sustainability Performance Targets:

- Net Carbon Footprint Upstream (Scope 1 and 2) equal to or lower than 5.2 MtonCO2eq (-65% vs 2018 baseline); and
- Renewable Installed Capacity equal to or greater than 5 GW.

The Bonds are expected to pay an annual coupon ranging between 2.625% and 3.125% but if one or both targets indicated above are not achieved, Eni shall pay an amount equal to 0.50% of the principal amount of the Bonds on the fourth interest payment date (14 September 2027).

Financial institutions involved

A number of banks and financial institutions were involved in providing support or guarantees to Eni's issued a sustainability-linked bond. These included:

- Global co-ordinator, documentation and facility agent – Mediobanca
- Global and sustainability co-ordinator – MUFG
- Global co-ordinators – Citi and Natixis
- Bookrunners – HSBC, UniCredit, and Intesa Sanpaolo
- Mandated lead arrangers – Bank of America, BNP Paribas, BPER Banca, Société Générale, and Wells Fargo
- Lead arrangers – Agricultural Bank of China, BBVA, Banco BPM, Barclays Bank, Credit Agricole CIB, Deutsche Bank, DNB Bank, First Abu Dhabi Bank, Goldman Sachs, J.P. Morgan, National Bank of Kuwait, Santander Corporate & Investment Banking, SMBC Bank, and Standard Chartered.

Source: (IEA, 2023[1]); (Eni, 2023[39]); (ESG News, 2023[40]).

Sustainability-linked bonds are not project specific, allowing for increased ability to balance strategies across portfolios. Similar to transition finance, the rationale for these products is consistent with the climate mitigation goal of reducing methane emissions in the oil and gas sector. However, the successful uptake of sustainability-linked bonds to address methane emissions in the oil and gas sector will depend on the existence of effective MRV mechanisms to preserve the integrity of the mechanism and avoid risks of greenwashing. Governments should be cognisant that these monitoring costs may be too high for all but large gas flaring and methane reduction projects (OECD, 2022[14]; World Bank, 2022[30]).

Debt for climate swaps

Debt-for-climate swaps – also referred to as debt-for-nature swaps – are financing mechanisms that may be able to help developing countries accelerate climate action by supporting projects that reduce methane emissions. In a debt-for-climate swap, bilateral creditors forgive host country debt and in return, the debtor government agrees to invest in national climate mitigation and adaptation projects, rather than continuing to make external payments to continue servicing its debt (OECD, 2022[14]; IGCD, 2020[41]). The rationale for these complex financial agreements is to provide emerging and developing economies with the fiscal space to respond to present environmental and climate-related challenges. The mechanism is particularly attractive if it can help unlock funding for climate projects, which do not ordinarily generate attractive financial returns but remain a priority for many emerging and developing economies (Jain and Verhoeven, 2023[42]).

Box 9.8. Debt for climate swaps in practice

Debt for climate swaps are a political and economic tool that seeks to address interrelated policy objectives – managing debt sustainability issues in often fiscally constrained developing countries; and catalysing climate action and sustainable development in developing countries.

Debt for climate swaps have three important elements: existing debt is repurchased at a discount through the use of a loan. Then, new bonds are issued at an interest rate which is below the market rate for the country and with a guarantee from a highly rated financial institution (which creates fiscal savings). Lastly, the debtor country commits to use part of those savings toward conservation activities (e.g. climate adaptation and biodiversity preservation).

The successful execution of a debt for climate swaps requires co-ordination by multiple actors (e.g. private creditors, governments, MDBs etc.). Such agreements ordinarily assume one of the following forms:

- **Two-Party (Bilateral) Swaps** – these are the simplest form of debt for climate swaps and involve the write-off of a bilateral loan by a creditor in return for agreed actions (generally environmental results within the debtor country). For example, in the 2006 TFCA swap for Botswana, the US Government forgave USD 8.3 million of bilateral debt in exchange for Botswana's commitment to facilitate grant financing for its tropical forests, funded partially through its savings from the swap.
- **Multi-Party (Commercial) Swaps** – these differ from bilateral swaps due to the involvement of third-party donors with the intention to buyout an existing sovereign debt instrument from the current creditors. In these agreements, the donor or intermediary will offer to purchase debt at a substantial discount from the creditors who currently hold it, and in return, the debtor country would allocate a portion of the savings to conservation activities – typically targeted to climate adaptation and biodiversity preservation. The primary intermediary in this arrangement is often an environmentally oriented NGO (e.g. CI, TNC or WWF) or a group of donor institutions may co-ordinate and share the burden of a single transaction.

Source: (Karaki and Bilal, 2023[43]); (Jain and Verhoeven, 2023[42]); (ANRC, 2022[44]).

Originally, debt-for-nature swaps were used for protecting and expanding carbon sinks and building resilience in ecosystems. In 1987, the first debt-for-climate swap was initiated by Bolivia, and subsequently, more than 100 similar transactions were facilitated during the 1980s and 1990s – before numbers peaked in1993 and a sharp decline in debt-for-climate swaps followed. However, in recent years, there has been resurgence of interest in debt-for-climate swaps, as an innovative finance mechanism, leading to the conclusion of new deals. These include the successful execution of debt-for-climate swaps in Seychelles in 2015 (blue bond issued in 2018), Belize in 2021, Barbados in 2022, and Ecuador and Gabon in 2023 (Jain and Verhoeven, 2023[42]). In addition, in 2022, debt swaps were explicitly referred to in the Sustainable Debt Coalition Initiative launched at the COP27 and featured at the 2023 Summit for a New Global Financial Pact among other innovative financial solutions to support developing countries, including those vulnerable to climate change (Karaki and Bilal, 2023[43]).

Debt-for-climate swaps could help developing countries accelerate climate action by strategically supporting projects that reduce methane emissions in their jurisdictions. Debt-for-climate swaps could provide fiscal space for indebted developing countries to invest in methane abatement measures including preventing flaring, venting and reducing fugitive emissions in the upstream oil and gas sector (IGCD, 2020[41]). However, this form of financial instrument may not be applicable in all country contexts

– especially in jurisdictions characterised by debt distress where debt restructuring should be prioritised in the first instance. Instead, debt-for-climate swaps could be utilised in context where debt level is significant but not unsustainable and where alternative financing instruments, including concessional finance and grants, are limited. In order for debt-for-climate swaps to be effective, they must be underpinned by a pipeline of interrelated projects and bankable projects to invest in, rather than one-off projects or transactions (Karaki and Bilal, 2023[43]).

However, while debt-for-climate swaps present an innovative opportunity for debtor nations to reduce their debt while making advances toward climate objectives, the widespread implementation of this mechanism is likely to be challenging. For example, although recent estimates suggest that climate-linked debt instruments could provide up to USD 105 billion of debt relief and help mobilise USD 329 billion in new borrowing, the total value of current debt-for-climate swaps is only a fraction of that amount (around USD 3.7 billion) and is nowhere near the USD 45 billion needed to achieve sufficient reductions in oil and gas methane emissions in low- and middle-income countries (Karaki and Bilal, 2023[43]; ANRC, 2022[44]; IEA, 2023[1]; IEA, 2024[2]). In addition, the capacity of some developing countries to manage large-scale climate mitigation projects, and the wisdom of insisting these governments spend resources on climate projects rather than provision of basic services, has also been questioned. Therefore, to overcome this objection and to attract the requisite creditor finance, debt-for-climate swaps should include criteria based on the Sustainable Development Goals to measure the impact on development gains to ensure that methane abatement measures not only reduce methane emissions, but also translate into improved energy access and the provision of services at the local level.

Notes

[1] While many SIFs are national funds (backed by governments), others are multilateral – for example, the IFC Global Infrastructure Fund, Emerging Africa Infrastructure Fund, Marguerite II.

References

ADB (2023), *2021 Energy Policy of the Asian Development Bank: Supporting Low-Carbon Transition in Asia and the Pacific*, Asian Development Bank, https://www.adb.org/documents/energy-policy-supporting-low-carbon-transition-asia-and-pacific. [17]

AfDB (2024), *The Ten-Year Strategy African Development Bank Group (2024 – 2033)*, African Development Bank Group, https://www.afdb.org/en/documents/ten-year-strategy-african-development-bank-group-2024-2033#:~:text=As%20the%20African%20Development%20Bank%20Group%20). [16]

AfDB (2012), *Energy Sector Policy of the AfDB Group*, African Development Bank Group, https://www.afdb.org/fileadmin/uploads/afdb/Documents/Policy-Documents/Energy_Sector_Policy_of_the_AfDB_Group.pdf. [15]

AfriCatalyst (2023), *Mobilizing Finance for Methane Action in Africa*, AfriCatalyst, https://africatalyst.com/wp-content/uploads/2024/01/Mobilizing-Finance-for-Methane-Action-in-Africa.pdf. [10]

AIIB (2022), *Energy Sector Strategy: Sustainable Energy for Tomorrow*, Asian Infrastructure Investment Bank, https://www.aiib.org/en/policies-strategies/strategies/sustainable-energy-asia/.content/index/_download/AIIB-Energy-Sector-Strategy-Update_Final_Nov-2022. [18]

Alberti, C. and B. Naran (2023), *Policy Brief: How to Start Scaling Methane Abatement Finance*, Climate Policy Initiative, https://www.climatepolicyinitiative.org/wp-content/uploads/2023/11/How-to-Start-Scaling-Methane-Abatement-Finance.pdf. [4]

ANRC (2022), *Debt-for-Nature-Swaps: Feasibility and Policy Significance in Africa's Natural Resources Sector*, African Natural Resources Management and Investment Centre, African Development Bank, https://www.greenpolicyplatform.org/sites/default/files/downloads/resource/debt-for-nature-swaps.pdf. [44]

CAF (2024), *Annual Report 2023*, CAF Development Bank of Latin America and the Caribbean, https://www.caf.com/media/4668518/annual-report-2024.pdf. [19]

CCAC (2023), *Côte d'Ivoire - Training workshops and guidance for methane leak detection and repair in the oil and gas sector*, Climate and Clean Air Coalition, https://www.ccacoalition.org/projects/cote-divoire-training-workshops-and-guidance-methane-leak-detection-and-repair-oil-and-gas-sector-0. [33]

CCAC (2023), *Gabon - Deliver a methane emissions inventory, mitigation plan and MRV framework for the oil and gas sector*, Climate and Clean Air Coalition, https://www.ccacoalition.org/projects/gabon-deliver-methane-emissions-inventory-mitigation-plan-and-mrv-framework-oil-and-gas-sector. [31]

CLDP (2023), *Methane Abatement Handbook*, U.S. Department of Commerce, https://cldp.doc.gov/sites/default/files/2023-09/CLDP%20Methane%20Abatement%20Handbook.pdf. [36]

Cordonnier, J. and D. Saygin (2023), "Financing solutions to foster industrial decarbonisation in emerging and developing economies", *OECD Environment Working Papers*, No. 226, OECD Publishing, Paris, https://doi.org/10.1787/24a155ab-en. [38]

de Aragão Fernandes, P. et al. (2023), *Landscape of Methane Abatement Finance 2023*, Climate Policy Initiative, https://www.climatepolicyinitiative.org/wp-content/uploads/2023/11/Landscape-of-Methane-Abatement-Finance.pdf. [3]

EBRD (2023), *Energy Sector Strategy 2024-28*, European Bank for Reconstruction and Development, https://www.ebrd.com/documents/power-and-energy/ebrd-energy-sector-strategy-20242028.pdf. [20]

EIB (2023), *EIB Energy Lending Policy: Supporting the Energy Transformation, Version with updated technical annexes - May 2023*, European Investment Bank, https://www.eib.org/attachments/lucalli/20230164_eib_energy_lending_policy_en.pdf. [21]

Eni (2023), *Eni launches the first Sustainability-linked convertible bond issue in its sector*, Eni S.p.A., https://www.eni.com/en-IT/media/press-release/2023/09/eni-launches-the-first-sustainability-linked-convertible-bond-issue-in-its-sector.html. [39]

ERA (2024), *Alberta announces $50 million from TIER Fund to reduce emissions, improve competitiveness through ERA's annual Industrial Transformation Challenge*, Emissions Reduction Alberta, https://www.eralberta.ca/media-releases/alberta-announces-50-million-from-tier-fund-to-reduce-emissions-improve-competitiveness-through-eras-annual-industrial-transformation-challenge/. [12]

ESG News (2023), *Eni Secures €3 Billion in Sustainability-Linked Credit for Greener Energy Goals*, ESG News Corp., https://esgnews.com/eni-secures-e3-billion-in-sustainability-linked-credit-for-greener-energy-goals/. [40]

GFMR (2023), *Global Flaring and Venting Regulations - Gabon*, Global Flaring and Methane Reduction Partnership (GFMR), https://flaringventingregulations.worldbank.org/gabon. [32]

Government of Alberta (2019), *Technology Innovation and Emissions Reduction Regulation*, Government of Alberta, https://open.alberta.ca/publications/2019_133. [11]

IDB (2020), *Environmental and Social Policy Framework*, Inter-American Development Bank, https://www.iadb.org/en/who-we-are/topics/environmental-and-social-solutions/environmental-and-social-policy-framework. [22]

IEA (2024), *Global Methane Tracker 2024*, International Energy Agency, Paris, https://www.iea.org/reports/global-methane-tracker-2024. [2]

IEA (2023), *Financing reductions in oil and gas methane emissions*, International Energy Agency, Paris, https://www.iea.org/reports/financing-reductions-in-oil-and-gas-methane-emissions. [1]

IEA (2021), *Financing Clean Energy Transitions in Emerging and Developing Economies*, International Energy Agency, Paris, https://www.iea.org/reports/financing-clean-energy-transitions-in-emerging-and-developing-economies. [13]

IFC (2021), *IFC Provides Landmark Loan to Reduce Gas Flaring, Boost Energy Access, and Power More Homes and Businesses Across Iraq*, Intenational Finance Corporation, https://www.ifc.org/en/pressroom/2021/ifc-provides-landmark-loan-to-reduce-gas-flaring-boost-energy-access-and-power-more-homes-and-businesses-across-iraq. [27]

IFC (2019), *AIMM Sector Framework Brief*, International Finance Corporation, https://www.ifc.org/content/dam/ifc/doc/2010/aimm-oil-and-gas-consultation.pdf. [26]

IFC (2017), *IFC's Definitions and Metrics for Climate- Related Activities (April 2017)*, International Finance Corporation, https://www.ifc.org/content/dam/ifc/doc/mgrt/ifc-climate-definitions-v31.pdf. [29]

IGCD (2020), *IGSD Background Note: Debt for Climate Swaps*, Institute for Governance & Sustainable Development, http://www.igsd.org/wp-content/uploads/2020/08/Background-Note-on-Debt-Swaps-11Aug20.pdf. [41]

IsDB (2019), *Sustainable Finance Framework*, Islamic Development Bank, https://www.isdb.org/sites/default/files/media/documents/2019-11/IsDB%20Sustainable%20Finance%20Framework%20(Nov%202019).pdf. [24]

IsDB (2018), *Energy Sector Policy: Sustainable Energy for Empowerment and Prosperity*, Islamic Development Bank, https://www.isdb.org/sectors/energy. [23]

Jain, L. and H. Verhoeven (2023), *Can Debt-for-Climate Swaps Help Heavily Indebted Developing Countries Address Climate Priorities?*, Center on Global Energy Policy, Colombia SIPA, https://www.energypolicy.columbia.edu/wp-content/uploads/2023/09/DebtClimateSwaps-Commentary_CGEP_090123.pdf. [42]

Karaki, K. and S. Bilal (2023), *Upscaling debt swaps for greater impact*, European Centre for Development Policy Management (ECDPM), https://ecdpm.org/work/upscaling-debt-swaps-greater-impact. [43]

Marcel, V. et al. (2023), *Left behind: emerging oil and gas producers in a warming world*, Climate Policy, https://doi.org/10.1080/14693062.2023.2231398. [6]

Mubadala Energy (2024), *Mubadala Energy*, Mubadala Energy, https://mubadalaenergy.com/. [34]

Mubadala Energy (2024), *Mubadala Energy - Portfolio*, Mubadala Energy, https://mubadalaenergy.com/portfolio/egypt/. [35]

Noels, J. and R. Jachnik (2022), "Assessing the climate consistency of finance: Taking stock of methodologies and their links to climate mitigation policy objectives", *OECD Environment Working Papers*, No. 200, OECD Publishing, Paris, https://doi.org/10.1787/d12005e7-en. [8]

OECD (2022), *Equitable Framework and Finance for Extractive-based Countries in Transition (EFFECT)*, OECD Development Policy Tools, OECD Publishing, Paris, https://doi.org/10.1787/7871c0ad-en. [14]

OECD (2022), *OECD Guidance on Transition Finance: Ensuring Credibility of Corporate Climate Transition Plans*, Green Finance and Investment, OECD Publishing, Paris, https://doi.org/10.1787/7c68a1ee-en. [7]

OECD (2020), *Developing Sustainable Finance Definitions and Taxonomies, Green Finance and Investment*, OECD Publishing, Paris, https://doi.org/10.1787/134a2dbe-en. [28]

OECD Development Centre (2020), *Working Toward an Inclusive and Equitable Framework for a Just Low-carbon Transition in Resource-Rich Developing Countries: Draft Work Plan 2021-22*, OECD Development Centre. [37]

UNFCCC (2015), *The Paris Agreement*, United Nations Framework Convention on Climate Change, https://unfccc.int/process-and-meetings/the-paris-agreement/the-paris-agreement. [5]

UNFCCC SCF (2021), *Biennial Assessment and Overview of Climate Finance Flows*, United Nations Framework Convention on Climate Change, https://unfccc.int/topics/climate-finance/resources/biennial-assessment-and-overview-ofclimate-finance-flows. [9]

World Bank (2022), *Financing Solutions to Reduce Natural Gas Flaring and Methane Emissions*, World Bank, http://hdl.handle.net/10986/37177. [30]

World Bank Group (2017), *World Bank Group Announcements at One Planet Summit*, World Bank Group, https://www.worldbank.org/en/news/press-release/2017/12/12/world-bank-group-announcements-at-one-planet-summit. [25]

www.ingramcontent.com/pod-product-compliance
Lightning Source LLC
Chambersburg PA
CBHW061223240926
51846CB00006B/1333
9789264567450